Editor: Liisa Janssens
Language editor: Floor Soesbergen
Design: Bob van Dijk, Petra Huijgens, Thiërry Tetenburg
Print: Akxifo, Poeldijk

ISBN 978 94 6298 449 3
e-ISBN 978 90 4853 515 6
DOI 10.5117/9789462984493
NUR 740

© Liisa Janssens / Amsterdam University Press B.V., Amsterdam 2016

Liisa Janssens (Ed.)

THE ART OF
ETHICS   IN THE
INFORMATION
SOCIETY

Amsterdam University Press, Amsterdam 2016

TABLE OF CONTENTS

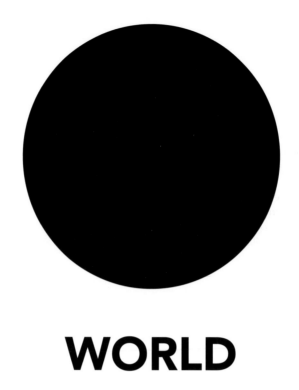

WORLD

PREFACE

We are very proud to present this essay collection *The Art of Ethics in the Information Society – Mind You*. The digital age is everywhere in our lives. It changes our jobs and employability. It disrupts existing models of trade, communication, care… life. It completely alters the range of actors, governors, and identities… Do we understand this? Do we want this? Can we stop this? Do we even want to stop this? These questions call for a visualisation of the ethical issues that are involved, since YOU, as a participant in the information society, need food for thought in this transitional phase. Food to be able to understand and decide, or even just to participate while preserving your own values and norms.

Since our start in 1999 the ambition of *ECP Platform for the Information Society,* is to establish a successful digital society that we can trust. ECP is leading, from a neutral non-profit position, in organising breakthrough programmes in the Netherlands. ECP and its members from business, science, government, and civil society foundations aim to stimulate the development of the information society, through creating the necessary economic, social, and political preconditions. A key element is to stimulate debate on ethics and technology; a debate that concerns the quality of life of all Dutch citizens as well as the economic competitiveness of the country.

The content of this essay collection can spark the imagination of YOU and everyone else in the digital society: OUR twenty-first century society. It can certainly help to provoke a discussion about the design of our society in the digital age.

We're saving you a seat in our FUTURE, just in case you want to be part of the JOURNEY,

Arie van Bellen

Director
ECP Platform for the Information Society

008

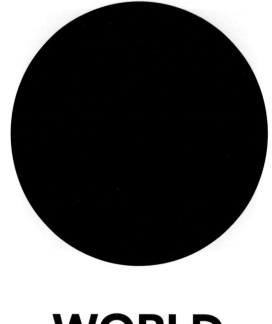

WORLD

INTRODUCTION

How do the numerous, and often rather alluring, promises of technology change our perspectives on ethical values? Specialists in the fields of science, art-science, and philosophy of technology share their forward-looking thoughts in this interdisciplinary essay collection. These visions, which provide new perspectives on values in the information society, present a unique view on metaethics and technology.

The contributions in this collection re-approach the ethical values of a reality in which the digital is increasingly merging, mostly invisibly, with the world around us. Technologies are often implemented prior to questioning their ethical consequences. We are all users and makers of the information society; therefore it is key to awaken everyone's imagination concerning ethics and technology. What are the ethical crossroads? The questions that are asked address this issue from multiple angles in order to show you the complexity of the different ethical choices that lay ahead of us. Central in this collection are *your* life in this rapidly digitalising world, and the matter of how we can shape the future with outlined choices in mind.

Can you imagine what your future life would look like if we were to continue implementing (new) technologies in our day-to-day lives at this pace? How do, and can, we shape our world through the possibilities offered by technologies? With the world going digital many questions concerning ethics arise: Will we acquire a perfect memory, and what are the consequences thereof? What will our future judicial system look like? Can computers make ethical decisions for us? Will it be normal to have a (sexual) relationship with a robot or with a digital personal assistant? Do we need to protect and conserve our cultural heritage digitally to ensure that time and war will no longer inflict damages? What role can the technologies that enable mixed reality play in end-of-life discussions? Will Big Data be Big Brother? Will the faith that we put in Big Data lead us to a Dystopia or a Utopia?

Knowledge of the future is vital for instigating and inspiring new and creative ideas to improve our lives. To design the ethical fundaments of the information society we need to use a futuristic approach in order to cope with the fast pace of the digital transition. The essayists wish to engage in a personal dialogue about the future with you, both as the user and maker, and it might be a dialogue that is beyond what you can foresee now. This collection can provide you with the knowledge on how you can take part in shaping the digital world. Since the secret to change is to focus all your energy on building the new, they hope to inspire you so that we can start building our future together.

Almost every successful person holds two beliefs: the future can be better than the present, and I have the power to make it so. In this endeavour, imagination of the ethical crossroads is a powerful tool; it is the source from which creativity springs. Artists can contribute to the reflections about the digital world and precarious technologies, scientists who look beyond the well-known possibilities can show us the challenges we are heading for, and philosophers with an open-minded reflection can awaken you through questions that we need to ask before technologies are implemented.

This collection revolves around you; you are challenged by visual and mental images, questions, and new ideas to reflect on your role during this transitional stage of a world that is moving into a digital era.

Do something today that your FUTURE will thank YOU for,

Go read this...

Liisa Janssens

BIOGRAPHIES

Nicky Assmann (1980) is a Dutch multidisciplinary artist, whose work is primarily perceived as a sensory experience, uniting the viewer with what lies before his eyes. A background in Film and Art Science combined with a keen interest in science and technology has lead Assmann to experiment with physical processes in the form of kinetic light installations, videos, and audiovisual performances. Her work has been exhibited amongst others at the Saatchi Gallery in London, at the Museum of Fine Arts of Taiwan, Woodstreet Galleries in Pittsburgh, Quartier 21 in Vienna, V2_Institute for the Unstable Media, and at Art Rotterdam Week. She holds a Bachelor of Arts in Media & Culture [Major Film Science] from the University of Amsterdam and a Master in ArtScience from the Interfaculty of the Royal Conservatoire & the Royal Academy of Art in The Hague. Assmann is a member of the art collective Macular, a collective research on art, science, technology, and perception. Since 2008 she has been part of the Sonic Acts curatorial team. She was nominated for the 2015 Prix de Rome and her work 'Solace' gained an Honorary Mention in the 2011 StartPoint Prize.

www.nickyassmann.net   www.macular.nl

Arie. J. M van Bellen (1962) is the founder director of ECP Platform for the Information Society. Since the foundation of ECP in 1999 he has been in charge of the organisation that employs 30 professionals who work on programmes and projects wherein business, government, society, and science collaborate to create the cornerstones of a digital society. He has a background in law (Leiden University, LLM) and bears the responsibility for national programmes regarding digital skills, cyber security, and visioning the impact of digitalisation on the quality of both economy and society. As director of ECP, he chairs various national committees, publishes works, leads debates, and is often a guest speaker on national conferences. Additionally, as a driving force behind the spreading of knowledge regarding the Information Society he, as director of ECP, has been involved as an impartial party in the design of national and international policies, such as the EU and the UN, regarding the digital agenda. Together with ECP he is the initiator of the ECP discussion group Ethical Aspects of the Information Society.

Tanne van Bree (1989) is a digital designer with a bachelor in Communication and Multimedia Design and a masters in Information Design (Mdes). She graduated Cum Laude from Design Academy Eindhoven with her graduation project, 'Artificial Ignorance'. She is interested in critical design, especially on the subject of technology. By designing 'digital forgetting' she speculates the different ways of dealing with the abundance of digital information available nowadays. Human memory is a duality of remembering and forgetting. This inspired 'Artificial Ignorance' – a computer application that offers a digital equivalent of 'forgetting'. Instead of displaying your digital photographs, AI collects visually similar images from the internet. These new images serve as 'memory cues' to stimulate active remembering.

www.tannevanbree.nl

Wouter Dammers (1985) graduated cum laude in 2009 from Tilburg University with a Master's degree in Law & Technology. He earned a second Master (with merits) from TU in International and European Public Law, part of this programme was completed at the Katholieke Universiteit Leuven, in Belgium, with a Erasmus scholarscip. In 2016 he intends to successfully complete the Grotius specialisation IT-law course. Before founding LAWFOX in 2013, he worked at SOLV, a law firm in Amsterdam, and legal consultancy firm ICTRecht, also in Amsterdam. With LAWFOX, Dammers is fully dedicated to managing conflicts relating to technology and law. LAWFOX is established in Tilburg, The Netherlands. Additionally, he has several legal publications to his name, including the section for 'The Netherlands' in The International Free and Open Source Software

Law Book. He is regularly asked by the press for his opinion on the latest developments in Internet law. Wouter Dammers is a co-author of the books 'ICTRecht - Cookies' and 'ICTRecht - Privacy'. Dammers also blogs regularly on his own weblog.

www.lawfox.nl

Frederik De Wilde (1975) works at the interstice of the art, science, and technology. The conceptual crux of his artistic praxis are the notions of the inaudible, intangible and invisible. De Wilde holds a master in Fine Arts and New Media, Art and Design (MA), and combined his studies in philosophy with a contemporary dance education. His art is often experimental and tries to offer new insights into the nature of art, science, and technology, how they interact (process) and how it can take shape (result). An excellent example is the conceptualisation, and creation, of the Blackest-Black art made in collaboration with American universities and NASA. The project received the Ars Electronica Next Idea Award and the Best European Collaboration Award between an artist and scientist, and was extensively covered (e.g. Huffington post, Creators Project, TED). In 2017 De Wilde will bring the Blackest-Black art to the Moon in collaboration with Carnegie Mellon, NASA, AstroRobotic, and Space-X. De Wilde collaborated with the KIT micro- and collective robotics lab, Rice University, Wyoming University, University Hasselt, KUL, and is both laureate and member of the Royal Belgian Academy of Sciences and Arts, a frequent speaker (e.g. TED ideas worth spreading), guest professor (e.g. interfaculty of art and science Amsterdam), publicist, and essayist (e.g. Experiencing the Unconventional, Science in Art). Currently De Wilde is developing several apps (e.g. VR and data visualisation apps), and is busy preparing his first short film as director and co-writer and is supported by the Flanders Audio visual Fund.

www.frederik-de-wilde.com

Bob van Dijk (1967) graduated cum laude at KABK (Royal Academy of Fine Arts, The Hague) and was immediately recruited by Studio Dumbar. Bob van Dijk was commissioned to work on the European side of the Euro coin. Two years later he was awarded the prestigious 'Dutch Design Prize' for his posters for 'Holland Dance Festival'. This campaign was recognised by a 'Typography Excellence Award' in New York and London. Van Dijk worked for clients like Telecom Italia, Hewlett Packard, NIKE, Coca Cola and SKODA. He cooperated with: Leagas Delaney - R/GA - Goodby, Silverstein & Partners - W+K - The Martin Agency - Ogilvy – Fallon. After van Dijk became Design Director at LAVA Amsterdam, they were awarded 'European Design Agency of the Year' that same year. The work of Bob van Dijk is part of the permanent collections of MOMA, Museum of Modern Art in New York and The Stedelijk Museum in Amsterdam. His work is published in several international books and magazines on graphic design and communication. Most recent and comprehensive is a fifteen- page interview in 'New Graphic Magazine' China. In 2006 van Dijk became a member of Alliance Graphique International (AGI).

www.bobvandijk.com

Rinie van Est and Lambèr Royakkers

Rinie van Est (1964) works as a research coordinator at the Rathenau Instituut, in the Netherlands. He is a physicist, a political scientist, and is specialised in the politics of innovation. He is primarily concerned with emerging technologies such as nanotechnology, cognitive sciences, persuasive technology, robotics, artificial intelligence, and synthetic biology. He has more than twenty years of experience in signaling trends in innovation and designing studies and debates about their meaning for society and democracy. He also lectures at the School of Innovation Sciences of the Eindhoven University of Technology. Some relevant publications: *Just ordinary robots: Automation from love to war (2016), Working on the robot society (2015), Intimate tech-*

nology: The battle for our body and behaviour (2014), From bio to NBIC: From medical practice to daily life (2014), Check in / check out: The public space as an Internet of Things (2011).

Lambèr Royakkers (1967) is the associate professor of Ethics and Technology at the Department School of Innovation Sciences of the Eindhoven University of Technology. Lambèr Royakkers has studied mathematics, philosophy, and law. In 1996, he obtained his PhD on the logic of legal norms. During the last few years, he has done research and published in the following areas: military ethics, robo-ethics, deontic logic and the moral responsibility in research networks. His research has an interdisciplinary character and focuses on the interface between ethics, law, and technology. In 2009, he started as project leader of the NWO-MVI research program: 'Moral fitness of military personnel in a networked operational environment' (2009 - 2014). He was involved in a European project, as chairman of the ethics advisory board of the FP7-project SUB-COP (SUicide Bomber COunteraction and Prevention, 2013 - 2016). Royakkers has authored and co-authored more than 10 books, including; Ethics, Engineering and Technology (Wiley-Blackwell, 2011), Moral Responsibility and the Problem of Many Hands (Taylor & Francis Gropup, 2015), and Just Ordinary Robots: Automation from love to War (CRC Press, 2016).

Jaap van den Herik and Cees de Laat

Jaap van den Herik (1947) studied mathematics at the Vrije Universiteit Amsterdam (with honours), received his PhD degree at Delft University of Technology in 1983 and was appointed as full Professor of Computer Science at Maastricht University in 1987. In 1988 he was appointed as part-time Professor of Law and Computer Science at the Leiden University. In 2008, he moved as Professor of Computer Science at the Tilburg University (2008 - 2016). In the Netherlands, he initiated the research area e-Humanities. Moreover, he was the supervisor of 71 PhD researchers. He was active in many organisations, such as JURIX (Honorary Chair), the BNVKI (Honorary Member), the CSVN (Honorary Member), the ICGA, NWO-NCF, ToKeN, CATCH, and the consortium BIGGRID. Van den Herik is ECCAI fellow since 2003, a member of the TWINS (the research council for sciences of the KNAW) and a member of the Royal Holland Society of Sciences and Humanities (KHMW). In 2012 he was co-recipient of an ERC Advanced Research Grant (together with Jos Vermaseren (PI, Nikhef) and Aske Plaat). On January 1, 2014, the appointment at the Faculty of Law was broadened to the Faculty of Science. Together with Joost Kok and Jacqueline Meulman he launched Leiden Centre of Data Science (LCDS) and is Chair of the Board of LCDS.

Cees de Laat (1956) chairs the System and Network Engineering (SNE) laboratory in the Informatics Institute of the Faculty of Science at The University of Amsterdam. The SNE lab conducts research on leading computer systems of all scales, ranging from global-scale systems and networks to embedded devices. Across these multiple scales, our particular interest is on extra-functional properties of systems, such as; performance, programmability, productivity, security, trust, sustainability, and, last but not least, the societal impact of emerging systems-related technologies. De Laat serves on the Lawrence Berkeley Laboratory Policy Board on matters regarding ESnet, he is also co-founder of the Global Lambda Integrated Facility (GLIF), the founder of GRIDforum.nl and the founding member of CineGrid.org. His group is/was part of the EU projects SWITCH, CYCLONE, ENVRIplus and ENVRI, EuroBrazil, Geysers, NOVI, NEXTGRID, EGEE, and others. He is a member of the Advisory Board Internet Society Netherlands and Scientific technical advisory board of SURF Netherlands.

delaat.net

Mireille Hildebrandt (1958) is a lawyer and a philosopher. She is a Tenured Research Professor of 'Interfacing Law and Technology' at Vrije Universiteit Brussels, where she works with the research group on Law, Science, Technology & Society studies (LSTS) of the Faculty of Law and Criminology. She also holds the Chair of 'Smart Environments, Data Protection and the Rule of Law' at the institute of Computing and Information Sciences (iCIS) of the Science Faculty at Radboud University Nijmegen. Hildebrandt has led research teams in numerous funded research projects, been editor, advisory board member and part of scientific committees and organised a number of conferences, workshops and seminars. She has given a number of keynotes on topics such as 'Slaves to Big Data. Or Are We?' (Barcelona 2013), 'Law as Information in the Era of Datadriven Agency' (London School of Economics 2015), and 'Learning as a Machine. Crossovers Between Humans and Machines' (Edinburgh 2016). She was part of the Onlife Initiative that published the Onlife Manifesto and is a founding member of the Digital Enlightenment Forum. She publishes extensively on the nexus of artificial intelligence, philosophy and law, for instance Profiling the European Citizen. Cross-Disciplinary Perspectives (Springer 2008), and the Routledge studies on The Philosophy of Law Meets the Philosophy of Technology (2011, 2013, 2016). Her most recent monograph is Smart Technologies and the End(s) of Law. Novel Entanglements of Law and Technology (Edward Elgar 2015).

works.bepress.com/mireille_hildebrandt

Liisa Janssens (1986) is Advisor Ethics at ECP Platform for the Information Society. Her responsibilities at ECP are introducing ethics and art by creating and initiating new programs on these subjects. She initiated projects such as a technology and art exhibition and a think-tank on ethics to constitute a multi-disciplinary and cross over cooperation between science, technology-art, philosophy of technology and the public and private sector. She obtained a Master's degree in Law at Leiden Law School (LLM) and a Master's degree in Philosophy at Leiden University (MA). She was assigned to the project on ethics and 'The Internet of Things' by the Dutch Cyber Security Council (National Coordinator for Security and Counterterrorism of the Dutch Ministry of Security and Justice). During her studies she was a trainee at SOLV, a law firm in the Netherlands that is fully dedicated to technology, media and communications, and she was a trainee at the Dutch Scientific Council for Government Policy (WRR) where she did research in the domain of the philosophy of technology and law for the book The Public Core of the Internet.

013

l.a.w.janssens@umail.leidenuniv.nl

Esther Keymolen (1982) is an assistant professor at Leiden University. Within the Law School, she works at eLaw, the Center for Law and Digital Technologies. She is the Academic Coordinator of the Advanced Master Law and Digital Technologies (LLM). Keymolen has research interests in the Philosophy of Technology, online trust, privacy, digital governance, and data ethics. Recently, her PhD thesis 'Trust on the Line. A Philosophical Exploration of Trust in the Networked Era' has been published by Wolf Legal Publisher. Previously, she worked at the Scientific Council for Government Policy (WRR). As a scientific staff member, she co-authored the book iGovernment and conducted research in the domain of digital youth care. In 2008, she completed her Master's degree in philosophy with a distinction and her Bachelor's degree with honours in 2007. She also holds a Bachelor's degree in Music (2004).

www.estherkeymolen.nl

Yolande Kolstee (1955) studied Social Sciences at Leiden University. In 2002 she started to work at the Royal Academy of Art, The Hague. She founded the AR (augmented reality) Lab in 2006. The AR Lab was a cooperation of the Academy, research groups of Delft, University of Technology and, after 2010, also with Media Technology from Leiden University. She and her team of students, teachers, artists, and scientists were part of new-media networks. In 2008 TUD developed the first optical see-through augmented reality headset. The AR Lab accomplished various projects with museums as Boijmans van Beuningen, Van Gogh Museum, and

Catharijne Convent as well as with SME. From 2011 to 2015 she held the post of Lector (Dutch for a researcher at universities of applied sciences (HBO) in the field of Innovative Visualisation Techniques in Art Education. Additional to her current position as officer for educational and international affairs at the Academy, she researches people's (lack of) interest in sustainability.

Marjolein Lanzing (1988) is a PhD candidate at the 3TU Centre for Ethics and Technology and works at the department of Philosophy and Ethics at the school of Innovation Sciences at Eindhoven University of Technology. Her PhD-project 'The Transparent Self: Identity and Relationships in a Digital Age' will contain a normative interpretation of the changing norms of privacy under the perspective of the changing meaning of the Self in a digital age. Her research entails an analysis of changing privacy norms ensuing from new ICTs and what this entails for the meaningfulness of self-relations and social relationships. Lanzing is the editorial assistant at the Philosophical Explorations, a peer-reviewed philosophy journal, specialising in the philosophy of mind and action and a member of the Amsterdam Platform for Privacy Research.

Ben van Lier (1957) is the director of Strategy & Innovation at Centric, a Dutch IT company with offices in Norway, Sweden, Germany, Belgium, and Romania. In this capacity, he focuses on research and analysis of developments in the interface between organisations and technology. Alongside his work at Centric, he obtained his PhD from the Rotterdam School of Management in 2009 (Erasmus University Rotterdam) on a doctoral thesis: 'Luhmann meets the Matrix: Exchanging and Sharing information within Network Centric Environments'. In 2013, he was appointed Professor at Steinbeis University Berlin. In this role, he focuses on qualitative research into topics such as systems and complexity theory, interoperability of information, and the network-centric approach. In 2015 he was also appointed Professor at the University of Applied Science Rotterdam, which focused on the development of the (industrial) Internet of Things. Both roles he fulfils next to his work at Centric.

014 **Koert van Mensvoort** (1974) is an artist, technologist, and philosopher who holds a MSc in computer sciences from Eindhoven University of Technology (1997) a MFA from the Sandberg Institute, Masters of Rietveld Academy, Amsterdam (2000) and a PhD in industrial design from Eindhoven University of Technology (2009). Van Mensvoort is best known for his work on the philosophical concept of Next Nature, which revolves around the idea that our technological environment has become so complex, omnipresent, and autonomous that it is best perceived as a nature of its own. It is his aim to better understand our co-evolutionary relationship with technology and help set out a track towards a future that is rewarding for both humankind and the planet at large. Among his works are the NANO Supermarket (a travelling exhibition disguised as a supermarket that presents speculative future technologies) the Datafountain (an internet enabled water fountain connected to money currency rates), the book *Next Nature: Nature changes along with us*, the Rayfish Footwear project (about a fictional company that creates bio-customised sneakers from genetically engineered stingray leather) and the In Vitro Meat Cookbook (exploring the potential impact of lab-grown meat on our food culture). Van Mensvoort directs the Next Nature Network in Amsterdam. He is a fellow at the Eindhoven University of Technology, board member of the Society of Arts at the Dutch Royal Academy of Sciences and gives presentations worldwide.

www.mensvoort.com   www.nextnature.net

Elize de Mul (1987) studied film and theatre studies (BA) and game studies and new media studies (MA) at the University of Utrecht and philosophy at the Erasmus University Rotterdam (MA). Her broad education reflects her fascination with a large range of things, mainly in the sphere of popular culture and the everyday life. This leads to a colourful assembly of topics and publications, amongst which a philosophical exploration of plastic bags published in book form (*Dansen met een plastic zak - Kleine filosofie van een onooglijk ding*, Klement 2014), a study of narrative identity in the television series *How I Met Your Mother,* and an attack on modern dualism using *Alice in Wonderland* as a philosophical tool. Currently, she works as a PhD candidate at the eLaw institute (Leiden University), where she researches the influence of the use of (popular) technologies on human identity and privacy, focusing on Internet memes, selfies, and quantified baby technologies. One day per week she is a lecturer of philosophy at ArtEZ, School of the Arts, where she tries to evoke a fascination and wonderment of 'the every day' in her students.

René P.H. Munnik (1952) studied chemistry, theology, and philosophy. His PhD thesis (1987) concerned the relation of mathematics, and metaphysics in the later works of Alfred North Whitehead. At the moment he owes the Thomas More chair of philosophy at the University of Twente and is senior lecturer in philosophical anthropology, hermeneutics, and metaphysics at Tilburg University, both in the Netherlands. Initially, his research was devoted to the interplay of natural sciences and humanities from a process perspective. See his: 'Whitehead's Hermeneutical Cosmology' (D.J.N. Middleton ed. *God, Literature and Process Thought,* Ashgate, Aldershot/Burlington 2002, p. 63-75). But gradually it evolved into research concerning the anthropological and cultural significance of technology. See his: 'Donna Haraway: Cyborgs for Earthly Survival' (H. Achterhuis ed. 2001. *American Philosophy of Technology: The Empirical Turn.* Indiana Univ. Press: Bloomington/Indianapolis, 95-118) and 'ICT and the Character of Finitude' (U. Görman, W. B. Drees, and H. Meisinger eds., 2005. *Creative Creatures. Values and Ethical Issues in Theology, Science and Technology.* T&T Clark: London/New York, 15-33). This contribution comprises parts of his Dutch book: *Tijdmachines* ('Time Machines'), mentioned in the bibliography.

Eric Parren (1983) is an interdisciplinary artist operating out of Los Angeles. His work lives at the intersection of art, science, and technology and investigates human connections to the ideas and technologies that shape our future. Eric Parren's works are often deeply sensory experiences dealing with modes of perception and the physics of light and sound. Through close study of the histories of media arts, electronic music composition, as well as abstract film, his work makes the link between the past, the present, and what is to come. Parren is a member of the art collective Macular and hosts the experimental music show La Force Sauvage on KCHUNG Radio. He is an instructor at Art Center College of Design where he teaches courses on programming for artists, virtual reality, and app development. Parren received his MFA from the Design Media Arts department at UCLA. His work has been shown at galleries and festivals across Europe, North America, and Asia since 2008.

www.ericparren.net

Marleen Postma (1989) is a graduate student in the MA Philosophy of the Humanities and the Research MA Literary Studies at the University of Amsterdam. She specialises in the relationship between ethics and literature and currently investigates how the influence of new technologies (such as lifelogging technology) may impact the way we live our lives and how it is represented in literature and the arts. Furthermore, she investigates, from a moral point of view, whether the act of reading literature can make us 'better people'. She is a member of the Brainwash Academy and has previously worked as a teaching assistant in the department of Media Studies and at the Institute of Interdisciplinary Studies of the University of Amsterdam, where she has contributed to the Big Data course.

Bart van der Sloot specialises in the area of Privacy and Big Data. He also publishes regularly on the liability of Internet Intermediaries, data protection and internet regulation. Key issues are the recently adopted General Data Protection Regulation, international data flows, especially between Europe and the United States, and data leaks. Bart van der Sloot has studied philosophy and law in the Netherlands, Italy and has also successfully completed the Honours Programme at the Radboud University. He is currently working at the Tilburg Institute for Law, Technology, and

Society at the University of Tilburg, Netherlands. Bart formerly worked at the Scientific Council for Government Policy (WRR) (part of the Prime Minister's Office of the Netherlands) to co-author a report on the regulation of Big Data in relation to security and privacy. In that context, he served as the first editor of a scientific book with contributions by leading international scholars and as the first author of an international comparative study on the regulation of Big Data.

bartvandersloot.com/index.html

Sabine Roeser (1970) is a professor of ethics and head of the Ethics and Philosophy of Technology Section of TU Delft, Netherlands. Sabine Roeser holds degrees in fine arts (BA), political science (MA) and philosophy (MA, PhD). Roeser has published numerous books and articles on risk, ethics, and emotions. She has obtained various competitive research grants. Roeser serves on various Dutch governmental advisory boards on risky technologies. Her research covers theoretical, foundational topics concerning the nature of moral knowledge, intuitions, emotions, and evaluative aspects of risk, but also urgent and hotly debated public issues on which her theoretical research can shed new light, such as nuclear energy, climate change, and public health issues. Recently she started studying the contribution that art can make to emotional-moral reflection on technological risks.

Floor Soesbergen language editor (1985) is an English teacher by profession and a storyteller by heart. She graduated from Utrecht University, with an MA in English Language and Literature (2008) and completed her second MA in Education (2011) after spending a year as an ESL teacher in Iran. As part of the Harting Programme she spent one academic year at the University of York (2006) where she specialised in metaphysical literature. She also completed a Pre-Master course at the University of Humanistic Studies (2014) and wrote her thesis titled 'The Art of Suffering - what can literature contribute to an ethics of humanism?' on bildung and personal growth achieved through reading and reflecting on Shakespeare's *Othello*. She enjoys finding and amplifying the unique sound of (academic) writers to strengthen the delivery of their message. Her passion lies in creating images and conveying stories and information that touch people's lives and stress a common humanity. In the weekends she can be found on various stages as a professional story teller.

Jelte Timmer (1986) is a researcher at the Rathenau Institute. His research discusses the social and ethical impacts of new and emerging information technologies, covering subjects from social networking, to smart mobility, big data, and persuasive technology. He has a background in psychology (Bsc), creative development (MA) and new media studies (MA). He has worked in the creative industry with Mediamatic in Amsterdam and he was one of the founding members of the Utrecht medialab- SETUP. Timmer combines the insights derived from psychology and cultural practices to reflect on societal-technological developments. Over the past years, he has published multiple books and essays on digital technologies. He has done studies on the ethics of digitisation for the Council of Europe and the Global summit of National Ethics Committees and brought together industry leaders at a Silicon Valley expert workshop on privacy in The Internet of Things. He has presented his work at international scientific conferences, industry meetings, and popular music festivals.

Maarten J. Verkerk (1953) studied chemistry and theoretical physics at the State University of Utrecht. In 1982 he earned a PhD from the University of Twente with a PhD thesis titled, 'Electrical Conductivity and Interface Properties of Oxygen Ion Conducting Materials'. In 2004 he earned a second PhD from the University of Maastricht with his thesis, 'Trust and Power on the Shop's Floor'. For several years he worked as a senior researcher in Philips's Laboratory for Physics in Eindhoven, the Netherlands. From 1986 to 2002 Verkerk was on several management teams and boards of various factories and developmental groups both in- and outside the Netherlands. From 2003 - 2007 he chaired the management board of psychiatric hospital 'Vijverdal' in Maastricht. Since 2008 he has been on the management board of VitaValley, an innovative network in healthcare. He has been an extraordinary professor in Reformational Philosophy at the University of Technology in Eindhoven since 2004 and at the University of Maastricht since 2008. He has published, amongst others, in the areas of materials science, feminism, business ethics, and philosophy of technology.

ABSTRACTS AND KEYWORDS

Nicky Assmann
Embodied Experience
The immaterial and intangible character of light, colour, and movement form the starting point of Nicky Assmann's spatial installations in which she tries to heighten the perception. In this essay she describes how she combines artistic, scientific, technological, and cinematographic knowledge in experiments with physical processes aimed at sensorial interference. She embeds her work in a context of visual music, expanded cinema, and the concept of synaesthesia. Set against the backdrop of our visual culture, where the perception of reality increasingly occurs in the virtual domain, she returns to the physical foundations of 'seeing' in which the embodied experience is central.
Keywords: art, science, perception, film, light, colour, movement, transition, sensorial, embodied experience, synaesthesia, technology, kinetic installation, video installation, visual music, expanded cinema, sublime, ephemeral, affect.

Tanne van Bree
Digital Hyperthymesia - On the Consequences of Living with Perfect Memory
Through the digitisation of the externalisation of human memory and a shift in cultural perspectives, a non-forgetting artificial memory evolves. In this essay Tanne van Bree uses a metaphor for this recently emerged phenomenon: she states that we are living with Digital Hyperthymesia. This is derived from the memory condition 'hyperthymesia', which gives a person a superior autobiographical memory, meaning that the person can recall, without conscious effort, nearly every day of their life with great detail. The emergence of Digital Hyperthymesia is researched from a technological and cultural perspective, and possible consequences in the context of human memory are formulated. Human memory consists of a duality of remembering and forgetting. This inspired experiments in designing a digital equivalent of forgetting, which resulted in Artificial Ignorance; a product aimed to counter the mentioned influences, and was intended to instigate debate on this subject.
Keywords: technology, digitisation, human memory, total recall, digital forgetting, algorithm, critical design.

Wouter Dammers
How to Govern King Code
With the convergence of technical revolutions, the regulating capabilities of code are becoming a threat to our human rights. As this new regulator is as non-transparent and uncontrollable as a metaphorical King in comparison to a democratic parliament, IT-attorney-at-law Wouter Dammers advocates to govern code. However, governance has its own issues, as our current democratic institutions struggle to keep up with technological advancements. Are we left to our (coded) fate? Or is our human intervention still necessary?
Keywords: governance, regulation, code, human rights, democracy.

Frederik De Wilde
Deep Learning Through Cinematic Space
This short essay is an attempt to question the power we attribute to technology by drawing potentially new connections between the use of Artificial Intelligences in Stanley Kubrick's *2001, A Space Odyssey* and contemporary usages of artificial intelligences like 'The Innovation Engine'. Kubrick gave us a warning. The subject of artificial intelligences, and their potentially far-reaching impact on society, will demand a moral and ethical compass and a global collaborative debate concerning the question how we want to evolve as a species.
Keywords: AI, evolution, art, *2001, A Space Odyssey*, Stanley Kubrick.

Rinie van Est and Lambèr Royakkers
Robotisation as Rationalisation – In Search for a Human Robot Future
This essay sees robotisation as a form of rationalisation. Driven by the belief in rationality and efficiency we have redesigned factories, offices and kitchens. Nowadays rationalisation touches every intimate aspect of our lives, from caring for the elderly to sex. Robots may contribute to this. Rationalisation is a double-edged phenomenon: besides benefits, it may reduce the freedom of people and lead to dehumanisation. The authors claim that robots can act as both humanising and dehumanising systems. They stress that even apparently typically human trades like face-to-face and skin-to-skin intimacy can eventually be lost to technology. Exactly because robots can have a profound effect on our humanity, we are in need of common moral principles and criteria for orienting ourselves into the robot future.
Keywords: robotisation, rationalisation, care robots, sex robots, dehumanisation.

Jaap van den Herik and Cees de Laat
The Future of Ethical Decisions Made by Computers
Moral issues and ethical decisions are usually seen as signs of civilisation. The idea that human beings and, in particular, human judges possess a privilege to judge upon ethical issues is widespread and ubiquitously accepted. However, the current development of disruptive technologies makes the following question acute: can computers outperform the best human judges in the area of moral issues? In this article, we discuss five fundamental problems (called invariants) with respect to the current state of the art. Our conclusion is that within two waves of disruptive developments (each taking, say, 25 years) computers will be on a par with, or even better in, taking ethical decisions than human beings.
Keywords: ethical decisions, privacy, security, invariants, disruptive technologies.

Mireille Hildebrandt
The New Imbroglio – Living with Machine Algorithms
This essay will discuss two types of algorithms: those capable of learning from their own 'mistakes' **019** and those that are not fitted with such capacities. The first concern machine learning (ML), the second can be categorised as 'dumb' 'if this then that' algorithms (IFTTTs). I will suggest that each can have added value as well as drawbacks, depending on how they are used, in what context and for what purpose. As the decision to engage either one of them may have far-reaching consequences for those subjected to their outcome, I propose that both should be constraint in ways that open them up to scrutiny and render their computational judgements liable to being nullified as a result of legal proceedings.
Keywords: machine learning, dumb algorithms, automation, autonomic, uncertainty, legal certainty, Rule of Law, public administration.

Liisa Janssens
Freedom and Data Profiling
What would happen if a set of algorithms could identify our behaviour and possibly even our thoughts? Such sets of algorithms would drastically change our interactions and relationships as human beings. Today, sets of algorithms can already generate profiles on the basis of data about an individual user. In doing so, an 'average you' is mapped; this process is called data profiling. In data profiling, however, lies the risk that human beings will only be seen through conceptions that follow from a system that is guided by probability analyses. At a time when data analyses are increasingly present in our day-to-day reality, we ought to reflect on the question as to whether human beings can completely be categorised on the basis of their data profile, or whether man as 'the Other' also contains a mystery that reaches further than these analyses can reach.
Keywords: algorithms, data profiling, probability analyses, freedom, meta-ethics, dystopia, Nicolai Berdyaev, Jacques Ellul, The Internet of Things.

Esther Keymolen

A Utopian Belief in Big Data

This essay aims to unravel the reason why policy makers –and others as well- persistently believe that Big Data will make the future completely knowable and consequently solve a myriad of societal problems. Based on insights deriving from the philosophy of technology, it will be argued that although human life is always 'under construction', nevertheless, there exists a Utopian longing for a final ground that contemporary technology should provide us with. This one-sided belief in the power of technology makes people blind for the unforeseen consequences technology may have. Technology, and more specifically Big Data, can only serve as a temporary shelter, which time and time again human beings will have to improve and alter. Moreover, this all-encompassing desire for certainty and safety is not as desirable as it may seem at first sight. After all, a life stripped from its complexity, may turn out to be a boring life.
Keywords: Big Data, Utopian belief, contingency, data-driven policy, philosophy of technology.

Yolande Kolstee

Digital Conservation of Cultural Heritage

Digital conservation of cultural heritage is important for safeguarding, research, education, preservation, and virtual exchange of historical artefacts. Future developments will be, for example, an increasing digitalisation of archives to make them accessible for research since their content is sometimes unknown. Various scanning methods make it possible to compare artworks to learn about the materials used and the construction of the artefacts. Techniques like Artificial Reality and Virtual Reality make it possible to annotate in an interactive way. By converting the physical material to a digital format and by making the results available through the Internet, we make our cultural heritage available to anyone, anywhere, whether it be solely for taking pleasure in an esthetical sense but also for study and research purposes. In doing so we can contribute to a world in transition in which sharing, co-creating, and shared responsibility will overcome exclusiveness and loss of identity caused by the inaccessibility of artefacts of our own or our neighbours' roots.

 Keywords: digitisation, cultural heritage, sharing, scanning, safe-guarding.

Marjolein Lanzing

Pasts, Presents, and Prophecies - On Your Life Story and the (Re)Collection and Future Use of Your Data

Big Data is a revolutionary promise that carries implications for our sense of self and identity development that we cannot yet fully grasp. Where our personal data goes and what traces it leaves on its way will slowly become more apparent. The information we share of ourselves could make our lives easier by eliminating choices. However, the digital identity that we create is one that we cannot easily shake and we should be aware of the images of ourselves that we create, like, and share. The information trail we leave behind fuels the predictive power of profiles. Can we still forgive, forget, be forgiven, and be forgotten in a digital society? And if not, what does that mean for our freedom of choice, sense of agency and our future selves?
Keywords: profiling, predictive analysis, forgetting, identity, agency.

Ben van Lier

Can Connected Machines Learn to Behave Ethically?

The rapid development of artificial intelligence, the huge volume of data available in 'the cloud', and machines and software's increasing capacity for learning have prompted an ever more widespread debate on the social consequences of these developments. Autonomous cars or the application of autonomous weapon systems that run based on self-learning software are deemed capable of making life-and-death decisions, leading to questions about whether we as human beings will be able to control this kind of intelligent and interconnected machines. It is still unclear which basic features could be exploited in shaping an autonomous moral status for these intelligent systems. For learning and intelligent machines to develop ethical cognition, feedback loops

would have to be inserted between the autonomous and intelligent systems.
Keywords: technology, software, artificial intelligence, information, ethics.

Koert van Mensvoort
Exploring the Twilight Area Between Person and Product
Anthropomorphobia is the fear of recognising human characteristics in non-human objects. The term is a hybrid of two Greek-derived words: *Anthropomorphic* means 'of human form' and *phobia* means 'fear'. Perfume bottles shaped like beautiful ladies, the Senseo coffeemaker shaped – subtle, but nonetheless – like a serving butler, and, of course, there are the robots, mowing grass, vacuuming living rooms, and even caring for elderly people with dementia. Today more and more products are designed to exhibit anthropomorphic – that is, human – behaviour. At the same time, as a consequence of increasing technological capabilities, people are being increasingly radically cultivated and turned into products. This essay explores the blurring of the boundary between people and products. My ultimate argument will be that we can use our relationship with anthropomorphobia as a guiding principle in our future evolution.
Keywords: anthropomorphism, humanoid robots, human enhancement, emerging technologies, ethics.

Elize de Mul
Living Together With a Green Dot – Being Together Alone in Times of Hyper-connection
Our communication with others nowadays often takes place via the various screens in our lives. Friends are reduced to merely a green dot on a screen, comfortingly telling us they are there, or to two green checkmarks that assure us the message we have sent out to the world has been read by friendly eyes. It is fascinating to state that these new forms of connection are examples of an on-going alienation by means of technologies; of us being alone, even when we are together. Instead, we could look at this 'living together with green dots' as an example of not an on-going alienation, but of a new form of being together typical for this era of 'hyper-connection'.
Keywords: together alone, Turkle, Shaviro, connected, network, individual, modern subject, Descartes, monads, Leibniz, hyper-connection, Facebook, communication, friendship, connectivity.

René Munnik
Technology and the End of History – From Time Capsules to Time Machines
The introduction of writing, especially the alphabet, marked the transition from (oral, mythical) pre-history to history, because it allowed the past to leave its own articulated messages. So, history – consisting of 'historical facts', both 'absent' and 'objectively real' – had a beginning. Contemporary I&C technologies substitute written records for the formal identity of data and algorithms. In doing so, they blur the distinction between absent past facts and their contemporary representations. They allow the on-demand presence of past facts that do not become 'history' anymore. Consequently, these technologies mark the end of history and the transition to a post-historical era.
Keywords: history, ICT, media, representation, writing.

Eric Parren
The Emerging Post-Anthropocene
With the rise of ubiquitous computing and the ever-increasing amount of sensors and processors that are being deployed, we are constructing a planetary scale cybernetic feedback system of computation. As our techniques for creating artificial intelligences become more sophisticated - and their implementations become more widespread in their applications- we are slowly handing over control of this system to a form of machine intelligence that is unbeknownst to us. Soon we might not be living in a world in which direct human activity is the primary impact on the Earth's geology and ecosystems, but instead in a world where the 'intelligences' that run our planetary computation system are the most influential factor.

Keywords: anthropocene, artificial intelligence, machine learning, cybernetic ecology, ubiquitous computing.

Marleen Postma
The Ethics of Lifelogging – 'The Entire History of You'
In this paper, Marleen Postma investigates how the ethical risks of lifelogging, the practice of capturing and storing the totality of one's experiences in a personal, searchable archive, are represented in the Black Mirror episode 'The Entire History of You'. She discovers that the episode, an artistic representation of lifelogging, clearly depicts the risks of pernicious memory and pernicious surveillance. Furthermore, the episode shows us the ways in which lifelogging has the potential to change us and our interpersonal relationships. In doing so, it encourages us to think about the ethical risks of lifelogging and possibly helps us to better understand these risks.
Keywords: lifelogging, pernicious memory, pernicious surveillance, privacy, personal information management system.

Bart van der Sloot
Privacy as a Tactic of Norm Evasion, or Why the Question as to the Value of Privacy is Fruitless
Privacy aims at avoiding norms, whether they be legal, societal, religious or personal. Privacy should not be regarded as a value in itself, but as a tactic of questioning, limiting and curtailing the absoluteness of values and norms. If this concept of privacy is accepted, it becomes clear why the meaning and value of privacy differs from person to person, culture to culture and epoch to epoch. In truth, it is the norms that vary; the desire for privacy is only as wide or small as the values imposed. It can also help to shed light on on-going privacy discussions. The 'nothing to hide' argument may be taken as an example. If you have nothing to hide, so the argument goes, why be afraid of control and surveillance? The reaction has often been to either argue that everyone has something to hide, or to stress why it is important for people to have privacy.

For example, it has been pointed out that people need individual privacy in order to flourish, to develop as an autonomous person or to allow for unfettered experimentation. This; however, is, in general, a rather weak argument. How, for example, has the mass surveillance activities by the NSA undermined the personal autonomy of an ordinary American or European citizen? Moreover, many feel that national security and the protection of life and limbs is simply more important than being able to experiment unfettered in private. The rhetorical question "Who needs privacy when you are dead?" is often asked. This essay will argue that there may be a stronger argument to make when the focus is turned around, namely not by looking at privacy as an independent value, which might outweigh or counter other interests, but as a doctrine which essence it is to limit and curtail the reach and weight of other values.
Keywords: privacy, value, intrinsic, norms, NSA.

Sabine Roeser
How Art Can Contribute to Ethical Reflection on Risky Technologies
Current debates about risky technologies such as robotics and biotechnology are usually quite heated and end up in stalemates, due to the scientific and moral complexities of these risks. In this essay Sabine Roeser examines the role that works of art can play in such debates. Recently, artists have become involved with risky technologies; this is what Roeser calls 'techno-art'. Roeser argues that, by prompting emotions, techno-art helps to make abstract problems concrete, explore new scenarios, challenge the imagination, and broaden narrow personal perspectives. She also discusses the possible challenges for techno-art, such as how to preserve the non-instrumental nature of art while playing a role in public debates, and how to do so in a meaningful way.
Keywords: risk, technology, art, emotion, ethics.

Jelte Timmer

Techno-Animism – When Technology Talks Back

In Spike Jonze's science fiction movie *Her*, the protagonist –Theodore– falls in love with his computer's operating system. It may seem a futuristic idea, but it might be closer than we think. Since the introduction of Siri, Apple's digital assistant, it has become increasingly normal to speak to your phone. This means a fundamental change in how we interact with technology. Machines become active participants in conversations. But while our technologies get to know us better and better, they also become more inscrutable and mystical to the average user. This leads to different relations with digital technologies in which a new kind of technological animism could become an important way for users to explain actions of their technological environments.

Keywords: human-technology interaction, interaction design, voice user interface, smart environment, ambient intelligence.

Maarten Verkerk

Design of Wisdom Coaches for End-Of-Life Discussions – Mixed Reality, Complexity, Morality, and Normativity

This essay discusses the idea of wisdom coaches for end-of-life dialogues. It is shown that end-of-life discussions are dominated by the medical-technological imperative that results in a wide-spread overtreatment of patients. It is argued that end-of-life dialogues have to revolve around the meaning of dying. The question is: how do we facilitate end-of-life dialogues? The idea of a mixed-reality wisdom coach that uses gaming principles is explored. The complexity of the design is understood and supported by the ontology as developed by Dutch philosopher Herman Dooyeweerd.

Keywords: wisdom coach, end-of-life, mixed reality, spirituality, meaning of dying, Herman Dooyeweerd.

EMBODIED EXPERIENCE

Nicky Assmann

I graduated from the Interfaculty of Royal Conservatory and the Royal Academy of Art in The Hague in 2011 with the title Master of Art Science. Thanks to my Film Science studies at the University of Amsterdam, I have gained a background in film, science, and art. Ever since then I have been working as an artist, and I am part of the 'Macular art collective', which focuses on art, technology, science, and perception.

The immaterial and intangible character of light, colour, and movement form the starting point of my spatial installations, with which I try to heighten the perception. I examine mental and physical perception by using light, mechanics, and abstract geometry to create natural and optical phenomena. My research combines scientific and cinematographic knowledge in experiments with physical processes aimed at sensory experiences. The visual propositions evoked in my installations demand a careful configuration of their constituent parts: space, light, object, materials, kinetic sequence, and viewer experience. Only with the right balance is it possible to achieve an immersive sensory experience. Set against the backdrop of our visual culture, where the perception of reality is increasingly present in the virtual domain, I return to the physical foundations of seeing in which the embodied experience is central, with a sublime and ephemeral character.

Many of my experiments involve the properties, behaviour, and aesthetics of the materials I use, for example, liquid soap, liquid crystals, and metals. This 'phenomenological' approach is expressed using different types of media such as video installations, audio-visual performances,

and kinetic light installations. A central question here is how does the human body, with all its senses, perceive and relate to these organic materials and natural phenomena?

Fluid Solids

When composing with materials and light, I employ physiological phenomena and natural transition processes. This can be seen in light effects (reflection, refraction), it shows how materials change when they are influenced by temperature and moisture (crystallisation) and the oxidation of metals.

Turbulence and iridescence, as seen in oil on water and rainbows, were central to my graduation project, 'Solace', and the following installation 'Solaris'. Both 'Solace' and 'Solaris' explore the mental process and physical activity of seeing, showing the transitions in moving liquid soap film, displaying fascinating abstract images and organic visual patterns. At regular intervals a handcrafted apparatus creates a soap film as a temporary spatial intervention. Precise lighting reveals the inner movement of the soap film and shows a turbulent choreography of iridescent colour and fluid motion. The visitor watches, as gravity teasingly plays with the membrane, until inevitably the fragile film bursts.

Observations of the natural world impact my work and its scale. I look for imagery that recalls the cosmos, e.g. planetary star systems, or the microscopic, like with the video installation 'Liquid Solid' in which we film the freezing process of a soap film.[1] The video shows the colourful soap in the liquid substance slowly sinking down in the film of soap, until a vacuum of a very thin layer of water remains, in which frozen crystals whirl around. Only at a very low temperature an accelerated freezing process occurs, during which ice crystals transform into fractal-like patterns. In contrasting the micro and macro, I see recurring images and patterns, which I use to zoom out and contemplate, and to obtain a sense of reflection and relativity.

Method and Objective

For every new work that I want to create I start by examining an elementary natural phenomenon, such as iridescence, and experimenting with associated materials, such as soap film. In my studio I experiment with liquids, materials, mechanical prototypes, and light set-ups. In recent years, I have focused on interdisciplinary installations incorporating light, mechanics, electronics, and software. Technology has enabled me to create energetic interplays of light and material, and I am also inspired by the long history of kinetic art dating from the 1920s by artists such as László Moholy-Nagy, Alexander Calder, Nicolas Schöffer, and Otto Piene.

The technology that I use is a means to an end, like a paintbrush to a painter. It has a supportive role and should not become the main focus of the work. The same goes for the natural phenomena or scientific research; these are merely sources of inspiration and they are not the objective. Since, I believe that, when a work of art aims to present (new) technology or simply visualises a phenomenon its timeless quality is reduced. I try to transcend both the phenomenon and the technology by creating a composition of multiple elements. Creating a composition of visuals and light, or composing visual music, is the main objective of the work.

When working on a composition, I allow myself to be guided by the material: light effects (such as reflection and light refraction), the way the materials change under the influence of temperature and moisture such as crystallisation, and the oxidation of metals. The material and the influencing factors are engaged in a multi-facetted dialogue, including movement, form, and the arrangement of space. I wish to create new imagery and unexpected moments by interrupting the natural dialogue of all these elements through various means of intervention.

My work refers to cinematography, and more specifically the concept of expanded cinema, which means to experiment with different elements of the cinematic apparatus: image, sound, space, and the embodied experience. My preference to visualise a natural phenomenon by using 'solid' materials is closely linked to my wish to incorporate the physical element and the presence of the physical body and its senses. Although the visual prevails, all senses are required to experience the work in its entirety: the embodied experience.

1 In collaboration with artist Joris Strijbos and filmed in the arctic region of Finland.

In 'Solace' I try to stimulate as many senses as possible: the subtle sound of bursting soap film, the scent of soap, and the skin's tactility, they play important roles in how the work is perceived.

The skin, whether visible or not, is a frequently recurring metaphor in my work, and this concept is extended into the space; coloured shadow play and reflections projected in the space shroud the visitor in multiple moving colours. The use of the fragile surfaces and thin membranes with mirroring qualities echo a distorted reality, as with the soap membrane of 'Solace', the transparent plates in the kinetic light installation 'Radiant', and the copper sheets in 'Aurora'.

026 · Aurora, 2015.

Patterns and Materiality

The analogue images I create often have a digital reference, such as with 'Solace', in which the monumental transparent screen is raised as an interruption of the space. It creates a temporary layer that can be seen as a visualised metaphorical reflection on augmented reality.

In recent work[2] I made use of rasterised lines in combination with moving light, this combination creates a moiré effect, which is strongly reminiscent of digitally created effects in computer graphics. In the kinetic light installation 'Radiant' undulating lines and dynamic coloured transparent acrylic plates generate the moiré patterns.

Materials and phenomena that also have qualities that refer to digital visual culture, such as digitally created effects in computer graphics, fascinate me. However, I have a need for a physical manifestation, for a certain 'materiality'. To me, materiality means the properties of materials, such as texture and tactility, which appeal to multiple senses and convey sensory messages. It creates, for me, a form of synaesthesia; the visitor can see what the material feels like. The material's texture emerges from the image without needing to touch it; it is merely a small tactile reference.

By letting go of the immaterial, or by re-materialising a digital image, I try to create a physical experience by building a bridge between digital imagery and a tactile interface. This provides an interesting combination and contrast of materiality and immateriality, and analogue and digital media.

2 The kinetic light machines 'Moiré Studies', in collaboration with artist Joris Strijbos.

027 · Moiré Studies, 2013-2016, The moiré effect, which is
strongly reminiscent of digitally created effects in computer graphics.

027 · Liquid Solid, 2015, Nicky Assman and Joris Strijbos, macro-level crystals.

DIGITAL HYPERTHYMESIA ON THE CONSEQUENCES OF LIVING WITH PERFECT MEMORY

Tanne van Bree

We are currently all living with Digital Hyperthymesia: a condition that enables us to have a perfect artificial memory. This non-forgetting memory is created by information stored externally and information uploaded to the Internet every day. We outsource part of our memories to this digital variant and it is affecting us.

As a digital designer I became intrigued by the inconspicuous influence certain everyday technologies have on us. Especially when I noticed that my English vocabulary seemed to have shrunk, which I traced back to a newfound Google Translate reliance. Therefore, I set out to investigate the results of technological development specifically on the subject of human memory. 'Digital Hyperthymesia' is a metaphor I created during my research to describe the phenomenon of the digitisation of the externalisation of human memory processes. It is derived from the condition Hyperthymesia: a rare neuropsychological condition characterised by a superior memory. The research resulted in a design product that aims to counter the mentioned influences. Additionally, it is meant to instigate debate on this subject.

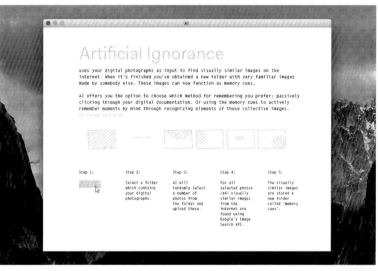

028 · Interface of computer application Artificial Ignorance.

The Case

In the paper 'A Case of Unusual Autobiographical Remembering' the authors propose the name hyperthymestic syndrome or hyperthymesia, from the Greek word *thymesis*, meaning 'remembering' and *hyper*, meaning 'more than normal'.[1] They report the case of AJ, 'a woman whose remembering dominates her life. Her memory is 'nonstop, uncontrollable, and automatic.' AJ spends an excessive amount of time recalling her personal past with considerable accuracy and reliability.'[2] People who suffer from hyperthymesia can recall almost every day of their lives nearly perfectly without conscious effort.

Such a powerful memory seems like a great advantage. Extraordinary memory capacity is an admirable and desirable trait. Furthermore, the activity of forgetting has had a negative connotation for centuries. However, the conclusions Parker, Cahill, and McGaugh reached on the case of AJ were that although 'her recollections were highly [...] accurate, and she obtained a perfect score on the Autobiographical Memory Test' her superior memory surprisingly 'does not necessarily facilitate other aspects of everyday life [...] and causes her to spend much of her time recollect-

1 By Elizabeth S. Parker, Larry Cahill and James L. McGaugh 2006.
2 Parker, Cahill, and McGaugh 2006: 35-49.

ing the past instead of orienting to the present and the future.' Also, 'neuropsychological tests documented that AJ, while of average intelligence, has significant deficits in executive functions involving abstraction, self-generated organisation and mental control.'[3]

The human memory consists of a duality: the activities of remembering and forgetting. While forgetting has a negative connotation, it is actually a valuable and essential function of human memory, and it allows it to work efficiently. In the case of AJ, a deficiency in forgetting resulted in dwelling on the past, not being able to orient on the present or future and having difficulty with conceptualising and exerting mental control. The constant, irrepressible stream of memories interrupts her everyday life, and has a destructive effect on her cognitive capacity. This strongly resembles the suggestion by Daniel Schacter that memory's fallibility ('sins') should not be viewed as a flaw. Instead, 'the seven sins reflect the operation otherwise adaptive features of memory.'[4] He argues that the ability to remember the gist of what happened is also one of the memory's strengths. This is 'fundamental to such abilities as categorisation and comprehension', allowing us to generalise across and organise our experiences. 'Noting that such generalisation "is central to our ability to act intelligently" and constitutes a foundation for cognitive development, McClelland contends that generalisation 'is central to our ability to act intelligently.'[5][6]

Therefore this paper made me question if a contemporary society in the 'information era' is subject to the same consequences of living with a shortage of forgetting as AJ, the woman with superior autobiographical memory? How did this phenomenon arise and will this have a similar destructive effect on our cognitive capacity?

Origin
We all live with Digital Hyperthymesia, a condition that enables us to have a perfect memory. It has become culturally accepted that recalling memories does not require much effort. We can simply explore our online or offline databases of memories by entering a word to describe the memory next to a digital magnifying glass, or we can search the right memory using filters (date, name, tags). In this essay I focus on self-created information, not on information gathered by other parties. Why is it that we are not hesitant to outsource an increasing amount of our memory to a digital one? And, what consequences can be expected through this development?

Technological Drivers
The first, and possibly the most obvious, cause of our recently acquired dependency on digital memory is the technological development that has made digital memory possible and affordable for everyone. In his book *Delete*, Viktor Mayer-Schönberger identifies four technological drivers that facilitate a shift in which remembering has become the norm and the exception: digitisation, cheap storage, easy retrieval, and global reach.[7][8]

Cultural Perspectives
The technological drivers played an important role in the evolution of Digital Hyperthymesia, this together with a gradual shift in cultural views regarding memory seem to be at the heart of this phenomenon.

The externalisation of human memory with a purpose of enhancing (collective) intelligence has been pursued throughout history. The invention of the alphabet, the activities of reading and writing, offered for the first time in human history the potential for learning one another's thoughts and ideas without having to engage in a direct conversation, and, therefore, this information surpassed the boundaries of time and space. People could build on each other's ideas across decades and centuries and create shared knowledge.

We now have seemed to hit the fast-forward button, when we will we reach a critical point of externalising too much of our memory? Moreover, would we even notice if we had? Since it is technologically possible to almost save anything and everything, it seems that this is the main reason to store everything: simply because it can be done; recording for recording's sake. This has

---

3 Parker, Cahill, and McGaugh 2006: 35-49.
4 Schacter 1999: 182-203.
5 Schacter 1999: 182-203.
6 McClelland 1995: 69-90.

7 Mayer-Schönberger 2009: 249.
8 These drivers are elaborately explained in my thesis (van Bree 2014). This essay focuses on the cultural motives.

given rise to new movements like 'lifeloggers', which consist of people who wear small devices that automatically capture various different aspects their lives. These recordings can be replayed, enabling a total recall scenario where no memory van be forgotten. Not only does this give rise to the incorrect assumption that more information always leads to better quality decisions, it also affects the way we view human memory.

Throughout history metaphors were used to explain human memory. These metaphors changed from biological interpretations of a 'living and evolving organism of knowledge' to a technical description using the latest terms for information storage technologies. From a 'library' or 'filing cabinet' to a 'photographic memory', and the most recently used metaphor: a 'computer'. This is not a constructive perspective because it undermines the typical human elements of the functionality of memory. It has led to a *misinterpretation* of the concept of human memory, and this has created a negative connotation that still remains today; this is a conflation of memory and storage. Most people see human memory as an untrustworthy, forgetful, adaptive thing, because it lacks the ability to preserve information indefinitely like computers do. This is all true, although instead of regarding the forgetfulness as a negative aspect it appears to be a valuable and necessary human activity. We should review the old, biological metaphors, relearn our premise regarding human memory and find new metaphors to communicate. For instance, Borges concludes in his story 'Funes the Memorious' that 'to think is to forget' and that, possibly, it is forgetting, not remembering, that is the essence of what makes us human.

We now relate memory to vast amounts of exact storage, and because human memory does not have this capacity, we have come to *value* our own internal memory less than a digital memory. Our prevailing conception, living in the 'Information Age', is that computers are better at this, and therefore it is best to use as many external memory aids as possible. Our devaluation of our memory is in contrast with the value the ancient Greeks placed on memory. Then, memory was the core of culture, and every literate person was practiced in mnemonic techniques, the art of remembering was seen as the art of thinking.

A theory by David Brooks called 'the outsourced brain' demonstrates the way society now relates to human memory:

> I had thought that the magic of the information age was that it allowed us to know more, but then I realized … that it allows us to know less. It provides us with external cognitive servants – silicon memory systems, collaborative online filters, consumer preference algorithms and networked knowledge. We can burden these servants and liberate ourselves.[9]

This choice of words illustrates the shift from seeing the value in possessing information from memory to seeing it as a 'burden', one we should be 'liberated' from. It seems that besides valuing our own memory less, the value of internal knowledge is also decreasing in our society. It is rather ironic that living in 'the information age' actually seems to cause us to know less. What will happen when the access is removed, a blackout in our memory? Joshua Foer has a very interesting view on this development:

> Our culture is an edifice built of externalized memories. Erudition evolved from possessing information internally to knowing how and where to find it in the labyrinth world of external memory. But as our culture has transformed from one that was fundamentally based on internal memories to one that is fundamentally based on memories stored outside the brain, what are the implications for ourselves and for our society? What we've gained is indisputable. But what have we traded away?[10]

Consequences
We have created an omnipresent, collective, easy accessible, external source of information but we have weakened, or at least modified, individual *intelligence*. The way we think is changing: instead of memorising knowledge, we now remember the location to find said knowledge. This is

9 Brooks 2007.
10 Foer 2011: 307.

a shift from actual ownership to mere access of information. What effects will this shift have on our intelligence? I consider it a realistic possibility that this shift will decrease the quality of our ability to reflect, our ability to reason, and our decision-making capacity.

With the current rapid technological developments an over-optimistic sentiment regarding technology seems to have taken hold of our society, and we do not take the time to reflect on the implications. Evgeny Morozov coined the term 'solutionism' to describe a mentality in which technology is the answer to all our problems, 'solutionism presumes rather than investigates the problems that it is trying to solve, reaching for the answer before the questions have been fully asked.'[11] Digital Hyperthymesia can have quite extensive consequences on our society if we fail to critically reflect on processes as they unfold. Morozov wonders:

> How will this amelioration orgy end? The fallibility of human memory is conquered too, as […] tracking devices record and store everything we do. No need to feel nostalgic, Proust-style, about the petite madeleines you devoured as a child; since that moment is surely stored somewhere in your smartphone, you can stop fantasising and simply rewind to it directly.[12]

The cultural perspectives regarding human memory in ancient Greece and now have changed extremely. I think the main difference is that the reliance on artificial memory has transformed our *active* method of remembering into a *passive* one. Remembering changed from making an effort to locate something in your internal memory and to revive the memory, to simply browsing an external memory, often without specifically knowing what we are searching for. This results in a disturbance of the dualism of human memory. Our devaluation of the human memory and knowledge seems to have resulted in a vicious circle: because of our fear of forgetting, caused by undervaluing our human memory, we now outsource an increasing amount of our memory processes, which results in a smaller internal memory and a decreased quality of thought. This will decrease our value of human memory, etcetera.

Next to these changes in the *way* we remember and value remembering, Digital Hyperthymesia could also influence our behaviour when people become more aware that they are living with a perfect, external memory. Viktor Mayer-Schönberger asks in *Delete*: 'if all our past activities, transgressions or not, are always present, how can we disentangle ourselves from them in our thinking and decision-making?'[13] When you view memories as the core of a person's identity, it is logical to assume that the possibility of preserving an abundance of digital memories is going to affect a person's identity, behaviour and choices. Thus, if people live with this in mind, it may even cause them to self-censor. Research by van Dijck indicates that the younger generation is already using digital photography as a tool for identity formation and communication, next to the primary goal of photography, preserving a moment and creating memory.[14]

Critical Design

After researching the emergence of Digital Hyperthymesia, the goal of my product was to design a counteraction to revert the consequences of this phenomenon on human memory. Besides neutralising the effects caused by Digital Hyperthymesia, other goals were to raise awareness and instigate reflection. Inspired by the dualism of human memory, I found the appropriate method for the rehabilitating of human memory required a digital equivalent for the activity of forgetting. I wanted to design digital forgetting with the purpose of stimulating human remembering.

Designing Digital Forgetting

How does the digital equivalent of forgetting look? First, I considered non-digital external memory, because analogue media do not generate the same problematic consequences as digital media in this case. Analogue media have a forgetting-mechanism ingrained through the decreasing of quality in case of copying or the passing of time. A digital copy is an exact replica of the original, and time has no influence on the decay of the digital object, only on the

11 Morozov 2013: 413.
12 Morozov 2013: 413.
13 Mayer-Schönberger 2009: 249.

14 van Dijck 2008: 57-76.

hardware where the digital object resides. There is an option of the aging of the software that reads the digital object, but this would result in an incompatibility error and a blue screen, not in gradations of digital decay.

Therefore, it seemed interesting to find out how gradations of digital decay could look, based on the concept of decreasing in quality as a form of forgetting. But after many small experiments the process stranded because I was creating visualisations of decay, and not making any structural changes. Through *The Enduring Ephemeral* I realised that I was trying to create something that goes against the very principles of the medium, because digital media is focussed on preservation. As Wendy Hui Kyong Chun mentions, 'The major characteristic of digital media is memory.'[15]

Therefore I concluded that my digital equivalent of forgetting should not be a form of digital decay or deletion, but rather a form of insufficient retrieval or limited recollection. It also is logical to not comprise the storing-stage of artificial memory, because that would make it obsolete to use for externalising information.

Design Proposal
The result of the design experiments is Artificial Ignorance, a computer application that offers a digital equivalent of 'forgetting'. Instead of displaying your digital photographs, AI collects visually similar images from the internet. These new images serve as 'memory cues' to stimulate active remembering.

Field of Design
The focus of digital media on remembering has given rise to a need for reflection on this topic and design solutions could contribute to address interesting alternative suggestions. Furthermore, it calls for counteractions to create more awareness. The urgency is also becoming apparent in the cultural interest: a collective request for ephemeral technologies seems to be emerging and the market is responding to this demand.

032 Besides the cultural interest, this theme is also important for information designers. IBM illustrates the complexity and the volume of information we're currently dealing with: 'every day, we create 2.5 quintillion bytes of data — so much that 90% of the data in the world today has been created in the last two years alone.'[16] There is simply too much of it. Information design focuses on making complicated and enormous amounts of information comprehensible by visually displaying it as an abstracted representation. Maybe next to finding new methods of effectively visualising information, a structural change is preferable. Instead of designing new visualisations of information from information, designing *volatile* information or digital forgetting mechanisms. I think this perspective is very interesting to be reflected upon in the information design field.

15 Chun 2008: 148-171.
16 IBM at *www-01.ibm.com/software/data/bigdata/what-is-big-data.html.*

References

Borges, J. 1942. *Funes, the memorious.* Buenos Aires: Editorial Sur. 107-115.

van Bree, T. 2014. *Evolving Digital Hyperthymesia.* Eindhoven, 43.

Brooks, D. 2007. *The Outsourced Brain.* The New York Times.
Accessed at *www.nytimes.com/2007/10/26/opinion/26brooks.html.*

Chun, W. 2008. *The enduring ephemeral, or the future is a memory.* The University of
Chicago. Critical Inquiry 3, 148-171.

van Dijck, J. 2008. *Digital photography: communication, identity, memory.*
University of Amsterdam. Visual Communication. 57-76.

Foer, Joshua. 2011. *Moonwalking with Einstein: the art and science of remembering
everything.* New York: Penguin Group. 307.

IBM. *What is big data.* Accessed at *www-01.ibm.com/software/data/bigdata/what-is-big-data.html.*

Mayer-Schönberger, V. 2009. *Delete: The virtue of forgetting in the digital age.* New Jersey:
Princeton University Press. 249.

McClelland, J. L. 1995. Constructive memory and memory distortions: A parallel-distributed
processing approach. In Daniel L. Schacter (Ed.) *Memory distortion: How minds,
brains and societies reconstruct the past.* Cambridge: Harvard University Press, 69-90.

Morozov, E. 2013. *To save everything, click here.* New York: Penguin Group. 413.

Parker, Elizabeth S., Cahill, L., and McGaugh, J, L. 2006. *A case of unusual autobiographical
remembering.* Psychology Press, Neurocase 12, 35-49.

Schacter, D. 1999. *The seven sins of memory.* Harvard University:
American Psychologist. 182-203.

HOW TO GOVERN KING CODE
Wouter Dammers

With the convergence of technical revolutions we are entering into a new phase of our information society. DRM and Black Box Trading showed us the downsides that code can have on our
human rights. With the emergence of the Internet of Things machines we will see the development of coding machines. Together with the emergence of 'smart contracts' code will become a
regulator. I believe that this regulator will not be a democratic parliament, but a non-transparent,
uncontrollable King. The TheDAO-hack of earlier this year showed us the dangers of 'code as
a regulator' very clearly. Therefore, I am of the opinion that code should be governed. This for
sure will have its own issues, as our current democratic institutions struggle to keep up with the
technological developments. Are we left to our (coded) fate? Or is our human intervention still
necessary?

The Rise of the Code
As an attorney-at-law specialised in IT and technology, I see many corporations and organisations
with varying degrees of awareness of the potentials threats of (new) technologies. With our smartphones, we have more computer power in our pocket than the Apollo 11 had on board on its
journey to the moon. Thanks to big data, Google can predict the arrival of a flu epidemic sooner
than medical experts, and market researchers have more faith in emotion recognition techniques
than consumer feedback on an advertisement's impact. Technology is becoming more involved in
what goes on between us, among us and even inside us. Technology is learning more and more
about us, and technology is becoming more like us. With the convergence of technological revolutions in nanotechnology, biotechnology, information technology and cognitive technology (the
so-called 'NBIC-convergence') we are entering into a new phase of our information society.[1]

034   One of the triggers behind these advances is code, or algorithms laid down in code. Essentially, code is the more-or-less intellectual creation of instructions in a computer programme. In
today's society algorithms, laid down in code, influence, lead, and even regulate our behaviour. In
fact, code functions the same as law in a certain way: it regulates our behaviour through architecture. It can allow us to do certain actions, while preventing us from performing others. However,
rather than resulting from a democratic process, code is most of the time the result commercial
considerations. And although code is fundamentally a human initiative, we are on the verge of
witnessing code creating code… It is no longer a case of man shaping technology, but also of
technology shaping man. I believe that these developments will lead to the demise of core values
such as free speech, privacy, and other freedoms, if we do not take action now. In this essay, I will
elaborate on why I hold this belief. To that regard, firstly, I will give a few existing examples of
code that can affect our fundamental rights. Secondly, I will discuss some current developments
with code that could lead to big issues in the light of 'code as a regulator'. After that, I will provide some insights in how code can be governed. Finally, I suggest the legislature to take immediate action.

Code that Affects Our Behaviour
Digital rights management, abbreviated as DRM, is a clear example of (commercialised) code that
regulates our behaviour. DRM provides various access control mechanisms to restrict the usage
of products, such as hardware and copyrighted works. Through DRM, the right holder can try to
control the use, modification and distribution of this software. Moreover, it enables systems to
enforce this policy. As such, DRM provides enormous possibilities for preventing the infringement
of intellectual property rights. However, DRM is often criticised since it can also prevent legitimate
use, and stifle innovation and competition. It is, in fact, a privatisation of law.
           Another example of behaviour regulating code is Black box trading, also known as algo-
rithmic trading, or algo trading. It encompasses trading systems that are heavily reliant on com-

1 Bainbridge; Rocco 2005.

plex mathematical formulas and high-speed computer programmes to determine trading strate-gies.[2] These trading strategies use electronic platforms to enter trading orders with an algorithm. The algorithm executes pre-programmed trading instructions. High Frequency Trade is a special-ised form of Black box trading. These strategies utilise computers that make elaborate decisions to initiate orders based on electronically received information. The decisions are made so rapidly that human traders are incapable of processing the information required to make the decision, let alone executing the decision. While many experts laud the benefits of these developments, others express their concerns. For instance, greater reliance on Black box trading brings a greater risk that system failure can result in business interruption as well. Also, companies can become playthings of speculators and technical problems, or security issues, can lead to a complete sys-tem breakdown - leading to a market crash. For instance, on August 1, 2012, the Knight Capital Group experienced a technology issues in their automated systems causing a loss of $440 million. Furthermore, on May 6, 2010, Black box trading contributed to volatility during the May 6 2010 so-called 'Flash Crash': The Dow Jones Industrial Averaged suddenly plunged about 600 points - only to recover those losses within minutes.[3] These examples briefly show which consequences human-instructed code has already had and continues to have on our own fundamental rights and lives. These examples are based on The Internet – of which its nature depends on its architecture, which is designed and is changeable. This is just the start; the impact of code will only increase. With the emergence of the NBIC-convergence, code itself will be programming code as well: machines will communicate with machines, machines will interact with machines, and machines will instruct machines. For instance, with the rise of the Internet of Things, devices will be able to exchange huge amounts of data with other devices connected through the IoT-network. This exchange is possible without any human interaction. A whole new ecosystem of communication will arise, where the connected devices no longer only share information - machines even produce and exchange information themselves. Humans will no longer be involved.[4] Thus, refrigerators will know that your milk is out of date, then they will connect to the local supermarket to purchase a new one, and the supermarket's drone will bring the milk to your doorstep; there is no need for human interaction. But what if the refrigerator is also connected to the database of your health insurance (perhaps even without you knowing) - and it is only filled with pizzas, burgers, and coke? What about your privacy? What about your freedom to choose?

Code As a Regulator

An interesting development, together with the emergence of the Internet of Things, is the developments with regard to 'smart contracts' and the 'blockchain'. Smart contracts are code instructions (protocols) that are able to facilitate, verify, and enforce the negotiation and perfor-mance of rights and obligations. It is a very interesting technology for the financial sector, since payment transactions can be done in milliseconds instead of the current three days.[5] However, it can be used in any sector where a 'trusted third party' is used for completing certain transac-tions (cadastre, notary, lawyers, bike rentals, and so on). Smart contracts can be made partially or fully self-executing, self-enforcing, or both. The contracts are actually nothing more but code: it defines certain situations in which defined actions will be triggered. These contracts are 'stored' in a distributed, shared ledger called the 'blockchain'. The blockchain consists of data-struc-ture blocks, with each block holding batches of individual transactions. Each block contains a timestamp and a hash of the prior block in the blockchain, linking the two. The linked blocks form a chain, thus giving the database its name. The distributed shared ledger records every transac-tion that has ever occurred in the blockchain. It is protected by a cryptography so powerful that it is, in practice, impossible to break the system. Moreover, whenever a new transaction occurs, the blockchain authenticates it across the network before the transaction is included in a next block in the chain. The information stored in the blockchain is, in fact, unchangeable code. A famous ex-ample of the blockchain is the Bitcoin – a kind of cryptocurrency. Another kind of cryptocurrency, Ethereum provides the possibility of storing smart contracts in the blockchain.[6]

By means of smart contracts, transactions (or triggering events for certain transactions)

2 Lims, Lemke: 2014.
3 Treanor 2015.
4 Van Lier 2013.

5 Cascarilla 2015.
6 See also ethereum.org.

can be automatically concluded or executed. Plus, of course, this will provide us with a tremendous amount of opportunities for innovation, since it has the possibility to cut out the middleman in any sector, but it can also bring undesired consequences to our fundamental rights, such as the right to privacy, the right to freedom of bodily integrity, and the right of freedom of expression and information. What if code, without any human interaction, can allow the one, and forbid the other, to enter a certain public area? What if code (read: machines) decide who is allowed to vote, and who is not. What if code (read: machines) decide who has the right to public healthcare, and who has not. And what if that code is not changeable by human interference any longer?

Thus, with Lessig, I believe that code will present the greatest threat to our ideals, as well as their greatest promise.[7] In studies regarding 'code as a regulator' certain patterns can be revealed with regard to regulation by code:

1. Regulation by code is rule-bound, in the sense that the code can only act in accordance with the code, and without instructions it can do nothing.
2. Regulation by code is non-transparent, since even experts cannot always understand the reasoning behind the code's choices.[8]
3. Regulation by code is fragile, since it can go from fully functional to completely broken within the flip of a switch; and,
4. Regulation by code is automatic, since programming can prevent or allow certain actions without any further intervision, while human-made law may need the interpretation by courts to do so.[9]

A good example that illustrates these patterns is the recent TheDAO-hack in the Ethereum Blockchain.[10] Ethereum is a kind of cryptocurrency, based on blockchain technology, which enables the use of smart contracts. TheDAO stands for 'decentralized autonomous organization'. Members of TheDAO can buy shares, and, according to the number of shares they have, can vote on matters. So it resembles a kind of corporation, but it is actually just a smart contract on the Ethereum Blockchain. A very smart hacker found a shortcoming in the 'splitDAO()'-function in the TheDAO-code: when a member exits the investment scheme, it supplied some of its own code with the transaction – by means of which TheDAO know how to transfer Ethereum, and this is necessary for making the pay-out. However, the function for the pay-out was recursive, meaning that it could be called upon a second time. By calling upon this 'splitDAO()' function again before finishing, the pay-out-process repeated itself, transferring more Ethereum coin. This was infinitely repeatable. So, ultimately, it drains all TheDAO's coins. And since the blockchain is unchangeable, a recovery of the stolen money was not possible.[11] This example shows that code is never found, but only ever made, we can build, architect and code a society to protect the values that we believe are fundamental, or one where these values will disappear. How such programming regulates our society depends on the choices made by the creator of the code.[12] However, there is no democratic process; most of the time, the choices are made with commercial interest in mind. A choice can be good, or a choice can be bad. More specifically, when code is created by code, no balancing standards (norms, values) might even be included. The consequences of the choices of coding can only be dealt with when the code allows us to do so. Even worse, codes might then be the only experts that have the knowledge of how the code is made, how the code (might) work, and how it could be intervened with. Issues are practically hardly traceable, reproducible or solvable.

Could governing the code prevent this hack? I believe not' governing code appears to be an impossible challenge. In past decades, governments, legislators and courts struggled to keep up with the rapidly changing technologies and uses.[13] 'Code changes quickly, user adoption more slowly, legal contracting and judicial adaptation to new technologies slower yet, and regulation through legislation slowest of all.'[14] Moreover, it is indeed non-transparent and, in fact, unpredictable, like a feudal regulator. Governmental regulations will not prevent the occurrence of hacks or 'bad code'. Since we are at the verge of code that will be shaping our society itself, I believe that it is at utmost importance to take action now to prevent the (further) demise of core values.

7 Lessig 2006.
8 Wagner 2005.
9 Grimmelmann 2005.
10 Siegel 2016.
11 Buterin 2016.
12 Wagner 2005.
13 Brown and Marsden 2013.
14 Brown and Marsden 2013.

Governing King Code

Thus, is it possible to govern 'King Code' in some other way? There are currently three approaches to code regulation: first, continued technological and market-led self-regulation; second, reintroduction of state-led regulation; and, third, multi-stakeholder co-regulation.[15]

The first approach, self-regulation will not be maintainable, since code is rule-bound, non-transparent, fragile, and automatic. Code is not neutral, so a balance of interests by a third party should be necessary in order to protect our fundamental rights. The second approach, reintroduction of state-led regulation, could at least reserve statutory powers to oversee self-regulation for the protection of such human rights. However, governmental regulation has serious legitimacy deficits. Since, it will lead to a widespread lobby to protect and introduce new barriers with legislative approval, with a continued exclusion of wider civil society. In the end, it will not be possible to maintain.

Nevertheless, I believe that regulating code through governmental intervention is (still) inevitable. Without it, every democratic legitimacy is traceless. We should not wait for a situation to be brought before courts to bring us any hope, since code will be changed more quickly than the case could be brought before court. In which regard, emphasis must be placed on regulators taking control of fast-moving tech markets early in their life cycles.[16] This is where I believe, in accordance with Brown and Marsden, the best option might be the third approach: a multi-stakeholder co-regulation (what is essentially pluralistic politics, with the government setting the table and inviting the stakeholders to it and guiding policy along the way), anchored in the function of code. They tested the development of what they call a 'unified framework for research into Internet regulation'.[17] This regulation should include principles for regulatory intervention that balances due process, effectiveness and efficiency, and respect for human rights. They refer to these principles as 'prosumer law',[18] in which user groups argue for representation in the closed business-government dialogue. The responsibility for developing this regulating code should lie with the firms who shaped the code. After the TheDao-hack, the Ethereum community was looking for various solutions. An interesting debate between interested parties followed after an initiative of Ethereum founder Vitalik Buterin. On the one hand, some defended the unchangeable nature of the blockchain ('pacta sunt servanda', underpinned by the first approach of self-regulation), on the other hand, others proposed an updated version of the blockchain that undid the hack ('intervention'). The users of the Ethereum Blockchain were given a choice: the users could vote for or against supporting the hard-fork. With a hard fork, a new version of the Ethereum Blockchain would be established, hoping that it would obtain enough backup by the necessary miners and wallets. This would not be without risks: it could lead to replay-attacks (routing transactions in the 'old' blockchain to a node in the 'new' blockchain, so that the transaction takes place in both blockchains, which could lead to chaos) and private wallets could be 'pickpocketed'. It ultimately led to a vote in support of the hard-fork. The hack was dealt with, but the unchangeable fundament of the blockchain was, critics believed, irretrievably damaged.

This case indeed shows that the patterns apply: regulation by code is rule-bound, in the sense that the code can only act in accordance with the code (pacta sunt servanda, agreements must be honoured); it is non-transparent, since the 'splitDAO'-function was not sufficiently understood (or thought through); it is fragile, since the complete Ethereum-system was on the edge of breaking down; and the regulation by code is automatic, since the TheDao-hack was allowed by the code. It also shows that, since code is not self-aware (yet?), human intervention is still necessary to execute the rule of law. The voters deemed it necessary to intervene. Therefore, the importance of human supervision cannot be overstated. The entire existence of our legal system is based on continually searching for inconsistencies and deciding whether or not they fit within our ideal legal framework. Yet human supervision in the context of debugging does not guarantee success. For programmers, debugging is one of their most difficult and time-consuming tasks, and does not always yield positive results. I go one step further and suggest applying the same principles of 'good governance' to this kind of multi-stakeholder co-regulation that the European Union has applied to itself. Including principles of good governance such as openness, participation,

15 Brown and Marsden 2013.
16 Wu 2011.
17 Brown and Marsden 2013.
18 Brown and Marsden 2013.

accountability, effectiveness, and coherence, one can define the way in which a power-assigned entity uses its power to regulate behaviour.[19] In order to make such regulation, and therefore the code, more legitimate, more democratic, it should (be mandatory to) adopt these principles of good governance. It is legislature's turn to set the table and to invite the stakeholders to it, guiding policy along the way, anchored in the function of the code, while taking account of these principles of good governance.

038

References

Bainbridge; R. 2005. *Managing Nano-Bio-Info-Cogno Innovations – Converging Technologies in Society.* Springer.

Buterin,V. 17 June 2016. *CRITICAL UPDATE Re: DAO Vulnerability. ethereum.org*

Brown, I. and Marsden, C. 2013. *Regulating Code.* MIT Press.

Cascarilla, C, G. 2015. *Bitcoin, Blockchain, and the Future of Financial Transactions.* CFA Institute.

Dammers, W. 2007. *Comitology in the Decision Making Process of the European Union.* Wolf Legal Publishers.

Grimmelmann, J. 2005. *Regulation by Software.* The Yale Law Journal. 114, 1719-58.

Lessig, L. 2006. *Code: Version 2.0.* The Perseus Books Group.

Lier van, B. 2013. *Can Machines Communicate? - The Internet of Things and Interoperability of Information.* Engineering Management Research, 2:1, 55-66.

Lins, G. and Lemke, T. 2014. *Soft Dollars and Other Trading Activities.* The New Financial Industry. Alabama Law Review, §2:30.

Siegel, D. 25 June 2016. *Understanding The DAO Attack.* Coindesk.com.

Stefik, M. 1999. *The Internet Edge: Social, Technical, and Legal Challenges for a Networked World.* Cambridge: MIT Press.

The Etherheum Project, *ethereum.org.*

Treanor, J. 2015. *The 2010 'flash crash': how it unfolded.* The Guardian.

Wagner, R. P. 2004. *On Software Regulation.* University of Pennsylvania Law School, Public Law Working Paper 57; Institute for Law & Econ Research, Paper 04-17.

Wu, T. 2011. *Agency Threats.* Duke Law Journal. 60, 1841.

DEEP LEARNING IN CINEMATIC SPACE

Frederik De Wilde

It can only be attributable to human error.

HAL 9000[1]

I am an artist working on the interstice of art, science, and technology. My art is grounded in the interaction between complex biological, societal, and technological systems. The indistinct, diffuse, 'fuzzy' area where biological and technology overlap and commingle is my favoured ground. In this essay I will explore the relationship between art, science, and technology within the context of Artificial Intelligence (AI). A leitmotiv is the enigmatic black monolith[2] and HAL 9000, the AI in Kubrick's deeply philosophical, spiritual, and allegorical film *2001, A Space Odyssey*. This film, about the nature of man and his relationship with technology, awoke my interest in AI. The film is concerned with topics such as: the mystery of the universe, existentialism, human evolution, technology, artificial intelligence, extra-terrestrial life, and with powers and forces beyond men's comprehension. The film follows a voyage to Jupiter after the discovery of an enigmatic black monolith that affects human evolution. We witness the interaction between a sentient computer, called HAL 9000, an acronym for Heuristically Programmed Algorithmic Computer, and the ship's astronaut crew. During their space flight, the astronauts discuss whether or not HAL has feelings. The movie comes to the conclusion that, if HAL has feelings, it is definitely the desire for power. When Dave Bowman, one of astronauts, finally tries to shut HAL down, it starts to sing a song that was the first thing it had learnt to say; a return to its unconscious childhood at the moment of death. Bowman eventually finds the monolith and fast-forwards human evolution.

Kubrick encouraged people to explore their own interpretations of the film, and refused to offer an explanation of 'what really happened' in the movie, preferring instead to let audiences embrace their own ideas and theories. In an interview Kubrick stated:

> You're free to speculate as you wish about the philosophical and allegorical meaning of the film—and such speculation is one indication that it has succeeded in gripping the audience at a deep level—but I don't want to spell out a verbal road map for 2001 that every viewer will feel obligated to pursue or else fear he's missed the point.[3]

I accept Kubrick's open invitation and reflect on HAL 9000 as a metaphorical warning for the power of technology.

AI and HAL 9000

HAL 9000 phrases his purpose as an AI quite nicely: 'I am putting myself to the fullest possible use', AIs are supposed to be of use to mankind, to be of service, so far so good. However, HAL adds a tricky sub clause 'which is all I think that any conscious entity can ever hope to do.' A computer programme as a *conscious* entity? However, if an AI is modelled by, and after, human behaviour, why would an AI not have a sense of consciousness and possess human flaws, such as being prone to addiction, or turn power hungry? AI systems do not have feelings and they do not know right from wrong: they only know what they are trained to do. If we train them to steal, to cheat, to disable, to destroy, that is what they will do. Hence, one can conclude that we have a large responsibility in 'training' AIs. The most fearful and awe provoking thought, however, is when an AI starts to design, or improve, its own software. Then AI will 'evolve' from a 'mere conscious' mirror image of our human psyche to an entity that can *create*. At this point it is not unthinkable that this type of AI will ultimately 'fuse' with the Internet of Things, machines, and/or plants. If we live to see technology crossing the border between the organic and the non-organic world, then, regard-

1 *2001: A Space Odyssey*. Directed by Stanley Kubrick, 1968.
Metro-Goldwyn-Mayer Inc..
2 Visually, best described as a large black slab out of one material.
3 dpk.io/kubrick 1968.

less of how science fiction this may yet sound, AIs operating in the realm of quantum computing is no longer in a galaxy far away. We can only speculate of what a deep impact AI will have on future societies here on Earth as well as in outer space, which brings us again to *2001, A Space Odyssey*. HAL 9000 proved to be a power-hungry liar when he attributed a broken antenna – which he had sabotaged himself - to 'a human error' in an attempt to take control of the ship. HAL can be seen as a metaphor for people, organisations, and societies that cannot admit their flaws; instead they hide behind the 'human error' excuse for what may be (weak) signals of systemic problems. In the worst cases, like HAL, such organisations and societies conspire against and condemn the accused or the victims, fearful that their own flaws may be exposed.

Another problematic issue is the increasing interdependency of the astronauts on HAL 9000. Using this to reflect on our contemporary society, one might wonder what the potential long-term impact of outsourcing specific tasks to AIs, DNNs, robots, and the likes, can have on mankind? Will we lose certain capacities like memorising, spatial recognition or motor skills? People have already begun, with or without conscious decision, to relinquish personal control over everyday decisions in favour of increasingly sophisticated algorithms. To what extent are we willing to let someone, or something, else take the helm? Especially if we do not know how a computer processes, or will come to process, information.

The Innovation Engine
In collaboration with scientists Jeff Clune[4] and Anh Nguyen[5] I created an artwork entitled 'The Innovation Engine', which questions and researches the obscurities of how a computer 'thinks' and 'sees.' The Innovation Engine consists of a touchscreen allowing visitors to navigate through, and explore, a deep neural network. Inspired by the central nervous systems of animals, machines have an artificial neural network: a computer algorithm. The webcam analyses, real-time, what it sees and what it has been 'taught' to detect. What is detected is visualised as highlighted artificial neurons. The audience can then browse through all the neural layers and gain insight into how a computer 'thinks' and 'sees'. A voice tells visitors what layer they are looking at and what is happening. A second screen presents a real-time updated slideshow of computer-generated and computer-encoded **041** images made possible through evolutionary algorithms, which are unrecognisable to humans, but that state-of-the-art cutting edge Convolutional Neural Networks (Deep Neural Networks trained on ImageNet) with $\geq$ 99.99% certainty to be a familiar object. We wish to demonstrate the potential limits and flaws of machine comprehension, especially how they 'see' the world, by hacking and misleading the artificial neural networks. The innovation engine researches the failure of machines and computers to simulate the human mind. More specifically, I was inspired by the inability of machines to accurately simulate the process of evolution due to their lack of processing power and other key functions required to run complex simulations. To conclude, 'The Innovation Engine' demonstrates how Artificial Intelligence and Deep Neural Networks are easily fooled, it is a dystopian reality when you realise that, for example, the military already relies on AI during missions. In 2014, the former director of both the CIA and NSA proclaimed, 'we kill people based on metadata' and it appears they have used 'a machine learning algorithms […] to try and rate each person's likelihood of being a terrorist.'[6] Computer scientists believe that all AI techniques that create decision boundaries between classes (e.g. SVMs,[7] deep neural networks, etc.), which are known as discriminative models, are subject to this 'fooling' phenomenon.

041 · Spiderweb_Rzl-Dzl-AI, The Innovation Engine.

The Next Monolith
One of the fundamental questions I want people to reflect on with The Innovation Engine is how we want to evolve as a species. It will demand collaborative, multi-, and interdisciplinary ecologies to discuss and tackle this complex subject with potentially far-reaching consequences. The arts can play a crucial role as a questioner, imaginer, and catalyst, as *2001 Space Odyssey* demonstrates. We are in need of a global debate; where we discuss and decide what shape our society should take next. The future of our species is increasingly designed in petri dishes and computer labs, but without moral and ethical compass we might lose our way or become extinct.

4 Jeff Clune is an assistant professor computer science at Wyoming University, Wyoming, USA. *jeffclune.com*.
5 Anh Nguyen is a Ph.D student in computer science at Wyoming University, Wyoming, USA.
6 *Arstechnica.co.uk* 2016.
7 Support Vector Machines are supervised learning models with associated learning algorithms that analyse data used for classification and regression analysis.

Although the theme music might sound ominous, there is no reason not to make use of our HALs or refrain from setting out in search of a monolith. We just might come to the conclusion that the road we are on is one of the alternative paths, which will open up new exciting doors to finding out where we came from but also where we are going.

Space.

042 · Tank_Rzl-Dzl-AI, The Innovation Engine.

References

2001: A Space Odyssey. Directed by Stanley Kubrick, 1968. Metro-Goldwyn-Mayer Inc..

Grothoff, C. and Porup, J.M. 2016. The NSA's SKYNET program may be killing thousands of innocent people. *Ars Technica UK*. 16 February at: *arstechnica.co.uk/security/2016/02/the-nsas-skynet-program-may-be-killing-thousands-of-innocent-people/.*

Stanley Kubrick interview with Playboy magazine. 1968. Accessed at: *dpk.io/kubrick.*

ROBOTISATION AS RATIONALISATION IN SEARCH FOR A HUMAN ROBOT FUTURE[1]

Rinie van Est and Lambèr Royakkers

The idea that robots in the future will take over the control of this planet from the human race is very much alive today. At the end of 2015, public figures like Elon Musk and Stephen Hawking stated that full artificial intelligence could spell the end of the human race. Bill Joy, nicknamed the 'Edison of the Internet', was one of the first to raise awareness for this existential threat. In 2000 he wrote a controversial article, 'Why the future does not need us', in which he warns for the dangers of uncontrolled intelligent systems, stating that we may actually destroy ourselves if we continue to further developing intelligent systems. That said, not everybody feels negatively towards a society dominated by robots. Some praise it as a next step in evolution and a victory for intelligent life. Transhumanist Ray Kurzweil happily foresees a future in which the world will belong to people who have reached a next level in intelligence (cyborgs) and to intelligent machines (robots), sadly the 'ordinary' unenhanced man has no place other than to function as a kind of pet.[2] Are these visions of super-humans, super-intelligent machines and the end of the human race merely wild speculations or should we take these kinds of predictions seriously?

Our pragmatic answer is that we should take the issue of human control over intelligent machines as well as the theme of robots as potential dehumanising technologies very seriously. This essay will focus on the latter theme, and claims that robots can act as both humanising and dehumanising systems. An example of the former, since 2004 robot jockeys are used to replace children at camel races in various Gulf States. These children were often kidnapped from nearby, poorer countries, such as Sudan and Pakistan, and treated very badly. In situations like this there may even be moral obligation to apply robotics. The challenge, of course, is to stimulate such humanising effects, like using robots to take over 'dirty, dull and dangerous' activities, and to prevent dehumanising effects, such as by not allowing robots to nurture our children or take care of our elderly. We will discuss the theme of dehumanisation from the perspective of rationalisation.

Rationality and Irrationality

> When human robots are found, mechanical robots cannot be far behind. Once people are reduced to a few robot-like actions, it is a relatively easy step to replace them with mechanical robots.[3]

At the beginning of the Twentieth Century, social theorist Max Weber (1864 – 1920) found that the modern Western world had become dominated by a belief in rationality. Weber saw the bureaucracy as the paradigm for the rationalisation process in his day. Belief in efficiency led to the redesign of the factory and labour. Engineers not only mechanised separate actions but aimed to design the factory as one 'great efficient machine'. Rationalisation, however, took place in many social practices. For example, also the kitchen became seen 'as a factory that converted input (groceries) into output (meals) by means of specific activities, technologies, and spatial distances.'[4] In a similar fashion, offices, airports, and cities were defined in terms of flows that could be designed and mechanised in an integrated manner.

Weber discussed rationalisation as a double-edged phenomenon. On the one hand, it could have many benefits, such as broader access to cheaper products and services with consistent quality. On the other hand, 'rational' systems can possess a variety of irrationalities, such as inefficiency, unpredictability, incalculability, and loss of control, for example; too many rules can render bureaucracies inefficient. Max Weber was most concerned about the so-called iron cage of rationality; the idea that an emphasis on rationalisation can reduce the freedom and the choices people have, and can lead to dehumanisation.

1 The argument in this paper draws on our book Just Ordinary Robots: Automation from Love to War (Royakkers and Van Est 2016). Just ordi-nary robots explores the social significance of the actual and intended use of robots in five domains: the home robot, the care robot, the use of drones in the city, the car robot and military drones.
2 Kurzweil 2005.
3 Ritzer 1983: 105.

Robots Caring for Humans

The notions of rationalisation and dehumanisation play a key role in the debate regarding care robots. Care robots are often depicted as the epitome of effective and efficient care. It is the ultimate rationalisation of a concept that can neither fully be measured nor captured in sensors and data. From an ethical perspective the hot potato is care without any human contact. The deployment of care robots may give rise to several ethical issues depending on whether robots play a role as: (1) companion for the care recipient, (2) cognitive assistant for the care recipient, and (3) (supporter of the) caregiver. Before we discuss the ethics, it is important to realise that the actual deployment of these types of robots is not expected in the short term.

A robot as companion can lead to misleading relationships, particularly since people have an innate tendency to attribute human traits, emotions, and intentions to non-human entities, such as robots. This so-called anthropomorphism tends to increase trust in robots, which can be both used and misused to persuade people to engage in certain actions. The robot as companion technology also raises controversial images of lonely elderly people who only have contact with animals or humanoid robots. The ethical concerns about the pet robot focus on the degree of human contact that such technology brings about. Sparrow and Sparrow describe care robots as 'simulacra' replacing real involving, complex, but rewarding social interaction.[5] Other authors are more positive; Borenstein and Pearson believe that the deployment of a robot, such as the seal robot Paro, can relieve feelings of loneliness and isolation.[6] A companion robot may also help isolated elderly people to keep up their skills of social interaction.

Cognitive assistance care robots may meet the need for senior citizens to live independently at home for a longer time. A robot can assist someone to remember appointments, to take medication or to eat on time, and can also ask the care recipient questions; such as do you have pain in your leg and then make a log-entry. The use of care assistant robots also raises questions. How pushy may a robot become, for example, in reminding someone to take medication? What if someone refuses to take the medication? This situation demonstrates the danger of paternalism or creating authoritarian care robots.

The use of robots as (supporters of) caregivers raises social issues relating to the human dignity of the care recipient. Sharkey and Sharkey believe that when robots take over tasks such as feeding and lifting, the care recipients may consider themselves to be objects.[7] Another drawback may be the reduction in human contact caused by the use of care robots. It is expected that the contact between care recipients and human caregivers will increasingly be mediated by technology. The unpleasant question underlying all of this is: how many minutes of face-to-face human contact is a care recipient entitled to receive each day?

It is important however to observe the choice of the care recipient. Some people might prefer a human caregiver, while others may prefer the support of robots, depending on which one gives them a greater sense of self-worth. Robots can thus be used to make people more independent or to motivate them to go out more often. The elderly may, for example, keep up their social contacts as they can go outside independently with the help of robots; robots here are used as technologies to combat loneliness. Equally, when deploying robots to assist people when showering or going to the toilet, the robots are the key to independence. Again, the manner in which robots are deployed and the tasks they carry out are both of crucial importance. The more control the care recipient has over the robot, the less likely he or she is to feel objectified by the care robot.

Robotisation as Rationalisation

Faith in rationalisation implies that efficiency, predictability, calculability, and control through substituting technology for human judgement present dominant cultural values.[8] In our era of big data, control often refers to digital control by means of algorithms. Rational systems aim for greater control over the uncertainties of life, in particular over people, who present a major source of uncertainty. One way to limit the dependence on people is to replace them with

4 Van den Boogaard 2010: 137.
5 Sparrow and Sparrow 2006.
6 Borenstein and Pearson 2010.
7 Sharkey and Sharkey 2012.
8 Cf. Ritzer 1983.

machines. After all, robots and computers are less rebellious than humans. Replacing humans by robots, therefore, is the ultimate form of rationalisation.

For example, in the course of the twentieth century manufacturers have robotised their car factories. That robotisation was preceded by the far-reaching rationalisation of the work practice; originally, craftsmen ruled the production process. Then dividing the work into many simple partial activities paved the way for its mechanisation and eventually made it possible for robots to enter the factory. Now robots are moving into society, a central question becomes: to what extent are we willing to rationalise our social practices in order to enable robots to play a role, or even replace humans, in these practices?

Independently of robotics, processes of rationalisation have accelerated over the last decades and become globalised. In *The McDonaldization of Society*, Ritzer argues that no aspect of people's life is immune to rationalisation anymore.[9] He sees the fast-food restaurant as the paradigm for the rationalisation of contemporary society. The entire food chain – from farm and factory to consumption – is geared toward efficiency. Nowadays many US citizens opt for dining at fast-food restaurants or eating microwavable food in front of the TV, this to the detriment of cooking from scratch and 'quality time' family meals. Ritzer argues that the fast-food culture has various dehumanising effects: the meatpacking industry creates 'inhuman work in inhumane conditions' and consumers are dehumanised as 'diners are reduced to automatons rushing through a meal with little gratification derived from the dining experience or from the food itself.'[10]

The history of the car factory illustrates that rationalisation of social practices can be a stepping-stone to robotisation. Namely, since robots have limited physical, social and moral capacities they can only work in a robot-friendly environment. For example, when people want to employ a vacuum-cleaning robot, they have to hide cables, remove deep-pile carpet or light-weight objects from the floor. This process of rationalising the living room is known as *roombarisation*, referring to the Roomba, the first vacuum-cleaning robot. Floridi argues that already for decades we are rapidly adapting both our physical and mental living space to the limited capacities of ICTs, including social robots.[11] To summarise: we seem to be rationalising each aspect of our lives, and often in an ICT- or robot-friendly way, so ICTs, like robots, may contribute to these processes of rationalisation.

Making Sex Mechanical

Even social relationships and sex have become rationalised, and indeed ICT is a major driver of this process. Technology is nestling itself within us and between us, it collects much information about us and can sometimes even operate like us, that is, mimicking the facets of our individual behaviour. In short, information technology has become 'intimate technology.'[12] The rationalisation of sociability is evident in 'rationalized online systems such as Facebook, where friendship is reduced to clicking an 'add' button and never needing to interact with that 'friend' on an individual basis ever again.'[13] Also, sex has undergone substantial rationalisation. Aoyoma, a relationship counsellor in Tokyo, believes Japan is experiencing 'a flight from human intimacy.[14] Many Japanese young people have lost interest in conventional relationships and sex, because they find it 'too troublesome' (in Japanese: *mendokuzai*). Replacing them with convenient technologies, such as virtual-reality girlfriends or sex robots, can reduce dependence on 'complicated' humans. Some people even believe that the future of relationships in Japan and the rest of the world will be largely technology driven.[15] In this scenario the rationalisation of sex with sex robots will ultimately make sex mechanical.

Levy only sees the advantages of sex with robots, since robots 'behave in ways that one finds empathetic, always being loyal and having a combination of social, emotional, and intellectual skills that far exceeds the characteristics likely to be found in a human friend.'[16] According to Levy, it is almost a moral imperative that we work to make these theoretical robotic companions a reality, because these robots could add so much love and happiness to our world. Unlike Levy, Turkle fears that the use of sex robots will result in de-socialisation and social

9 Ritzer 2013.
10 Ritzer 2013: 135.
11 Floridi 2014.
12 Van Est 2014.
13 Flusty 2009: 436.
14 Haworth 2013.
15 Haworth 2013.
16 Levy 2007: 107.

de-skilling.[17] Sex will become purely a physical act without commitment rather than a loving and caring act. She describes a trend towards rejecting authentic human relationships for sociable, human-like robots, and wonders what kind of people we are becoming as we develop increasingly intimate relationships with machines. Other authors respond that Levy ignores the deep and nuanced notions of love and the concord of true friendship. Sullins argues that in the way sex robots are currently evolving 'we have an engineering scheme that would only satisfy, but not truly satisfy, our physical and emotional needs, while doing nothing for our moral growth.'[18] Also Richardson who leads a *Campaign against sex robots*, wants to raise awareness of this design issue and persuade those developing sex robots to rethink how their technology is used. She believes that they reinforce traditional stereotypes of women, objectification of women, and the view that a relationship need be nothing more than physical: 'these robots will contribute to gendered inequalities found in the sex industry.'[19]

A New Beginning

Robots can have a profound effect on how we define our relationships and ourselves with other human beings. Moreover, robots can act as both humanising and dehumanising systems. It is important to realise that the way robotics will develop and used in society is not fixed. There are many technical, economic, individual, social, and political choices to be made. Echoing the views of Weber and Ritzer, this essay acknowledges that modern society is obsessed with rationality. Now robotics is moving into society, we are challenged to weigh the potential social gains of rationalisation through robotisation against potential social costs.

Weber's label of the 'irrationality of rationality' comprises all the negative aspects and effects of rationalisation. It warns us that robotisation may lead to systems that become anti-human or even destructive of human beings and communities. In the 1980s, Ritzer observed that the rationalisation of food consumption caused the loss of the communal meal for which families got together every day. He regretted that because he found that family meals could play an important role in keeping families together. Ritzer argued:

> There is much talk these days about the disintegration of the family, and the fast-food restaurant may well be a crucial contributor to that disintegration. Conversely, the decline of the family creates ready-made customers for fast-food restaurants.[20]

Nowadays, various people fear that robotisation may foster similar types of vicious circles. For example, some would regret if companion care robots were to lead to a loss of interaction with real humans, or if sex robots would cause people to become physically and socially disconnected from each other, and that this in turn, would stimulate the demand for sex robots. It is important to acknowledge the possibility that apparently typical human trades, like face-to-face and skin-to-skin intimacy, can eventually be lost to technology. Precisely, because robots can have a profound effect on our humanity, we need to clarify through reflection and joint debate what kind of human qualities we do not want to lose. The scenarios about robots and (de)humanisation can stimulate that debate. So, instead of worrying too much about the end of the human race, let us begin searching for common moral principles and criteria for orienting ourselves into the robot future.

17 Turkle 2011.
18 Sullins 2012: 408.
19 Richardson 2015.
20 Ritzer 2013: 137.

References

Borenstein, J. and Pearson, Y. 2010. *Robot caregivers: harbingers of expanded freedom for all?* Ethics and Information Technology. 12:3, 277-288.

Floridi, L. 2014. *The Fourth Revolution: How the infosphere is reshaping human reality.* Oxford: Oxford University Press.

Flusty, S. 2009. *A review of "The McDonaldization of Society 5".* Annals of the Association of American Geographers. 99: 2, 435-437.

Haworth, A. 2013. Why have young people in Japan stopped having sex? *The Guardian / The Observer,* 20 October 2013.

Joy, B. 2000. Why the future doesn't need us. *Wired.* 8 April.

Kurzweil, R. 2005. *The Singularity is Near: When Humans Transcend Biology.* New York: Viking.

Levy, D. 2007. *Love + Sex with Robots. The Evolution of Human-Robot Relationships.* New York: HarperCollins Publishers.

Richardson, K. 2015. The asymmetrical 'relationship': Parallels between prostitution and the development of sex robots. *SIGCAS Computers and Society.* 45:3, 290-293.

Ritzer, G. 1983. *The McDonaldization of society.* Journal of American Culture. 6:1, 100-107.

Ritzer, G. 2013. *The McDonaldization of Society – 20th anniversary edition.* Thousand Oaks, CA: Pine Forge Press.

Royakkers. L.M.M. and Van Est, Q. 2016. *Just Ordinary Robots: Automation from Love to War.* Boca Raton, FL: CRC Press.

Sharkey, A. and Sharkey, N. 2012. *Granny and the robots: Ethical issues in robot care for the elderly.* Ethics and Information Technology. 14:1, 27-40.

Sparrow, R. and Linda Sparrow, L. 2006. *In the hands of machines? The future of aged care.* Mind and Machines. 16:2, 141-161.

Sullins, J.P. 2012. *Robots, love and sex: The ethics of building love machines.* Affective Computing. 3:4, 398-409.

Turkle, S. 2011. *Alone Together. Why we expect more from technology and less from each other.* New York: Basic Books.

Van den Boogaard, A. 2010. Site-specific innovation: The design of kitchens, offices, airports, and cities. *Technology and the Making of the Netherlands: The Age of Contested* Modernization, 1880-1970, J. Schot, H. Lintsen and A. Rip (eds.). Cambridge: MIT Press, 124-177.

Van Est, R. with assistance of V. Rerimassie, I. van Keulen and G. Dorren. 2014. *Intimate Technology: The battle for our body and behaviour.* The Hague: Rathenau Institute.

THE FUTURE OF ETHICAL DECISIONS MADE BY COMPUTERS

Jaap van den Herik and Cees de Laat[1]

The issue of using computers for judging court cases is a logical follow up of the question: Can computers play chess? This question dominated Artificial Intelligence research from 1950 to 1997, when DEEP BLUE defeated Kasparov.[2] For the true scientist, the remainder from this breakthrough was the research question on chess intuition.[3] Is intuition Programmable?[4] Of course, chess is a finite game and, in contrast, decisions on legal cases are a matter of infinity (uncountable, infinite). Here, the reader has to take into account that 1046 is the number of chess positions,[5] which still is seen as 'infinite' by the current high-performance computers. One complicating factor is that, with chess, the good outcome is universal (checkmate), however in court cases there is always a fuzziness in the outcome, and what is considered good judgement in one culture might be completely the opposite in another.

The Development

Computer science, its applications and its perverse threats to society have been among us for about seventy years. The start of the development can be set at 1945. Up to 1980 technological developments were leading. From 1980 to 2010, attention arose for the threats that computers pose to society. In the Netherlands, one of the first devil's advocates was Bob Herschberg.[6] In 1978, he gave his inaugural address *In de Ban van de Fout*.[7] In the 1980s he stimulated his students to engage in 'ethical hacking' to show society how vulnerable computers were and how easy it was to bypass security and access private files, and this vulnerability inspired an array of ethical questions.

From 2010 till now we can see the beginning of the first phase of the development of many disruptive technologies. An illustrative example of this development is taken from the PhD thesis *Measuring and Predicting Anonymity*.[8] The manuscript shows that combining data from many different – social media and Big Data – sources immensely affects the privacy of people and provides a mathematical framework for calculating that effect. After receiving his PhD Koot encountered a case regarding jobseekers; the case concerns a clever student of the University of Amsterdam. A gifted and competent entrepreneur had developed a test as to answer the question 'how fit am I to be an entrepreneur,' aimed at the unemployed who wished to start an enterprise. This clever student gave the ten-euro test, provided via a national bank's website, a try. A thorough analysis of the given answers (30 pages) was given to the client. Of course, this was also done online by providing the client a link, where he[9] could download the report. The link had the identification (anonymised by the authors) psychology.test.bank.entrepreneur.id=45210. Given that the website advertised with the phrase '45000 people have already taken the test' the question came up what would happen if the number 45210 were changed into 45209. Surprisingly, the results of the previous client appeared. Tests from roughly 3000 earlier IDs showed that the website had no security measures or rate limiters installed. The organisations involved (company and bank) were immediately informed using a responsible disclosure approach and the leaked data was encrypted and later destroyed.

The supervisor of this student immediately understood what had happened, and could foresee the, short and long term, consequences. He informed the Faculty and University Authorities and some of these did not understand the severity and expressed their criticism, they might have been afraid of legal claims or conflicts with the bank. Admittedly, the question was even more complex since one of the well-known banking institutes of the Netherlands, which encourages people to lend money for starting a company, had hired a one-person IT company for designing and administrating such a test. So, the consequences were best described as disastrous. However, everyone became very pleased after they understood that the supervisor and student had followed a responsible disclosure process as implemented by the research group involved.

1 The authors would like to thank the members of TWINS, an advice council of the Royal Academy of Science and Arts for fruitful discussions on this topic and also for the organisation of discussion meetings. In particular, we are grateful to Frank den Hollander, Arie Korbijn, Jan Bergstra, and Hans Dijkman.
2 Seirawan 1997.
3 Cf. De Groot 1946, 1965.
4 Cf. Van den Herik 2016.
5 Chinchalkar 1996.
6 From Delft University of Technology.
7 Herschberg 1978.
8 Koot 2012.
9 For brevity, we use 'he' and 'his' whenever 'he or she' and 'his or her' is meant.

Establishing an Ethical Board

The example of the psychological test was one in a string of similar encounters with technological vulnerabilities or outright life-threatening cases encountered by that supervisor and his colleagues. This shows that the art of computer science has become ridden with ethical issues, much like the medicinal sciences. To cope with the situation, and not ad hoc but in a structural way, the computer-science faculty collected information about procedures for different ethical boards in other sciences and came up with a structure and procedure for the University of Amsterdam – Informatics Institute. The Ethical Board consists of a number of senior staff members and a law advisor. The board reports to the director of the Informatics Institute, and all research proposals and student-research projects that have potential ethical issues have to pass the board. To aid the process there is also a subcommittee that handles the student research projects and there is an online question-naire that has to be completed by Principal Investigators (PIs) of research proposals.[10] This is similar to the current gender issues and research data-management plans that are also becoming manda-tory in EU and Science Foundation project proposals. The ECIS board is deemed so essential and successful that CWI and VU have joined the effort.

A very important action is to embed ethical behaviour in research in the earliest education that students receive. For example, the students of the master System and Network Engineering at the UvA receive lectures about responsible disclosure, anonymisation, and ethics in the first week of their study. Recently, the Royal Netherlands Academy of Arts and Sciences has emphasised the need for such education by their publication Ethical and Legal Aspects of Informatics Research.[11]

Invariants for the Development

The entrepreneurial case teaches us several lessons; below we mention four of them. The first lesson is on moral issues and ethical decisions. A person who is curious as to whether the before men-tioned testing company has done a decent job in (a) analysing the answers and (b) protecting the interests of other clients, is facing a variety of ethical questions. To what extent may he be interested in the test results by others; (i) to inspect whether test results are well protected? (ii) To see them? (iii) To read them? (iv) To retrieve two thousand of them? The answer to (i) given by Herschberg and his students in the 1980s was: inspecting whether data leaks are well prevented; it was considered a task of education. If not, then the company was informed and in some cases (e.g., if the company denied cooperation) the press (radio and TV) was alerted. Times have changed, but the questions are still prominent. The lesson learned is that less-honourable students might be seduced to make severe mistakes.

The second lesson teaches us about privacy. In the entrepreneurial case, it is clear that privacy protection was not taken into consideration at all. The liable persons are the company owner and, possibly, the bank. Each of them has learned their own lesson although the issue of privacy is far from being resolved at this moment. However, in such cases the reputation damage might be much higher than the liability.

The third lesson involves security. Every system designer should address the issue of secu-rity, defined as 'protect the system as well as possible (preferably in a perfect way) against intended damage'. The lesson is that security should be performed by design, not as an afterthought, and certainly not by obfuscation.

The fourth lesson concerns integrity. To what extent can (may/must) we expect integrity of the general public and computer users in particular? This is a difficult societal issue, best formulated as: what are the exact rules? Still, nowadays we have difficulties with defining the precise rules for data integrity.

Additional to these four issues that can be seen as fundamental problems in computer science, there is a fifth lesson to be learned (not applicable to the entrepreneurial case). This lesson is about safety. The issue is whether a computer system can be regarded as safe. Here we define safety as to 'protect the system as well as possible (preferably in a perfect way) against unintended damage'. Examples are: damaged caused by fire, a deluge of water, nuclear material, and earth-

10 See website ECIS.
11 Royal Netherlands Academy of Arts and Sciences 2016.

quakes. So, the development that started in the 1980s has led to five fundamental problem areas, which we call the *invariants* of our research.

New Concepts for Measuring the Invariants

The identification of five invariants sheds a clear light on the threats that can affect the functioning of a society. Obviously, the invariants do neither have dichotomous values, nor can they be ranked by numerical evaluations (only). Even if we assume that some ranking on actions, events, and issues should be possible, then the prevailing question still is: "what are the instruments to measure the invariants?" Not surprisingly, the instruments (tools and concepts) saw a rapid developed between 1980 and 2010.

Moreover, in the last six years we have seen the emergence of disruptive technologies. We mention as such Internet of Things, robotics, Big Data mining, autonomously driven cars, block-chains, and drones. All of them are cleverly thought-out concepts, but they also have unexpected effects. For now, we will use the concepts as they are meant and in a way that they can help us measuring and ranking the 'values' of the five invariants (see Table 1).

Five Invariants	Measured in 2016 with the help of
Ethical Issues	Responsibility
Privacy	Reciprocity
Security	Adaptivity
Safety	Autonomy
Integrity	Curation

051 • Table 1: Five Invariants and their current measures.

Since space is pressing, we provide only a short explanation of the invariants. The term 'ethical issues' has been superseded by the notion of responsibility. It 'measures' to what extent somebody or something is responsible (think of responsible data).

Some will say that privacy is dead.[12] We do not discuss whether this is true or slightly exaggerated. The main point we make is that a total protection of privacy is impossible (and maybe even undesirable). Therefore, at least in the public domain reciprocity should be enforced, so that people know that they have passed 1200 cameras during their trip from London city to their suburban home.

Security is a discipline on its own, and in the Netherlands Fox-IT plays a key role in this area. The most important points are adaptivity of the prevention and pro-active behaviour. This branch of science relies heavily on collecting knowledge, gathering information, and remaining up to date. This goes for all areas, ranging from psychology to artificial intelligence to high-performance computing.

Safety is closely related to security. Here the emphasis is on robustness and vision on what may be expected in the 'other' (non-computer) world. The main technique is that management in disasters should be cooperatively and dynamically executed in intelligent networks.[13] The question here is: to what extent do we give the system autonomy to implement the suggested adaptations in order to let the system survive?

Integrity is a human characteristic, but it is also used in computer science in the form of integrity of data. The key tactic to get data to meet the conditions for integrity is scrubbing, usually formulated as 'curation of data'. The reverse way from data to human beings is a possibility for returning to human integrity.

12 Cf. Morgan 2014.
13 See Weber 2017.

An Example of Criminal Behaviour

In order to show the subordinate position of the weights of the invariants, ethical decisions, and privacy, we focus on the investigation of criminal behaviour. We provide an example from the marathon bombing in Boston. In April 2014 an unexpected bomb exploded at the end of the Boston Marathon. There were sensor images and recorded pictures, but the crowd was so large and dense that no criminals involved in the attack could be identified at first inspection. The police quarantined the city and attempted to find the criminals in their computer databases. They could not identify them, since they missed clues that can be gathered from heuristic search. After much work, one brother was killed by police force (Dzjochar Tsjarnajev) in a gunfight and one day later Tamerlan Tsjarnajev was arrested, albeit by coincidence through a blood trace observed by an alert citizen.

To investigate such criminal behaviour with modern technology we have learned that a combination of Big Data, High Performance Computing, and Deep Learning is most effective. These three components form the basis of *Narrative Science*.[14] It means that an intelligent program constructs the story and points to the criminals. This approach is also known as 'story telling', while lawyers call it 'argumentation theory'.

In retrospect, it worked also in Boston, which implied that the criminals could be identified afterwards with the help of the identification stories, constructed by narrative science (see Figure 1).

The 'scientific' development of 'disruptive' technology, i.e., the Road to, among others, Deep Learning is as follows. After that we provide two other developments.

Artificial Intelligence	1950-1990
Machine Learning	1990-2000
Adaptivity	2000-2005
Dimension Reduction	2005-2010
Deep Learning	2010-2015
Big Data & HPC	2012-2017
New Statistics	2014-2019

If we now try to add a third column to Table 1 headed by 'measured in 2066 with the help of', then we will certainly find somewhere Deep Learning as a key notion to achieve the goal, e.g., for arriving at an ethical decision.

The Future

The future can only be predicted with adequate accuracy when we have a thorough knowledge of the history,[15] as well as the current scientific developments.[16] In general, we see two developments: (1) the dependency on intelligent computer programs is increasing, this holds for government, industry, companies, banking, and ordinary people; (2) the trust in computers and internet is diminishing, caused by hacking, fishing, internal interest setting (libor interest), and pollution setting (automobiles); the two curves are shown in Figure 2.

Figure 2 shows that the curves are divergent after 2010. We marked the distance between them for the year 2016 and argue that the discrepancy can be resolved by Privacy Enhancing Technologies (PETs).

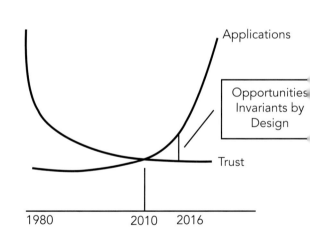

052 • Figure 2: Dependency of applications versus trust in the outcome of application

14 Cf. LeCun, Bengio, and Hinton 2015.
15 See e.g. van den Herik 1991; Hamburg 2007, Christensen 2009,
Adriaanse 2015.
16 Cf. van de Voort, Pieters, and Consoli 2015.

Measures

Having read about the complexities and intricacies of the five invariants, the main question is: what is our opinion on Disruptive Technologies, such as Big Data, High Performance Computing, and Deep Learning? To sharpen the two positions in such a discussion, we discern two possible directions: (1) to diminish attention and research efforts for Disruptive Technologies and (2) to increase attention and research efforts for Disruptive Technologies.

From these two directions our preference goes to the second direction, provided that we are allowed to introduce three specific safeguards. These are our *recommendations*:

1. Increase research on AI systems with Disruptive Technologies and emphasise on moral constraints.
2. Increase research on security of AI systems that are candidates to be hacked. In particular, use Privacy Enhancing Technologies.
3. Establish (a) a committee of Data Authorities and (b) an Ethical Committee.

After the implementation of these three recommendations we may conclude that within two waves of disruptive developments (each taking, say, 25 years) computers will be at a par with, or even better in, taking ethical decisions than human beings.

053

References

Adriaanse, J. 2015. Ethical Challenges in Turnaround Management. Lecture at *Leadership Challenges with Big Data. Turning Data into Business*, Erasmus University Rotterdam, June 30.

Chinchalkar, S. 1996. An Upper Bound for the Number of Reachable Positions. *ICCA Journal* 19:3, 181-183.

Christensen, B. 2009. *Can Robots Make Ethical Decisions?* Accessed at *www.livescience.com/5729-robots-ethical-decisions.html.*

ECIS, 'Ethical Code' at UvA: *ivi.uva.nl/research/ethical-code/ethical-code.html.*

de Groot, A.D. 1946. *Het denken van den schaker, een experimenteel-psychologische studie.* PhD thesis, University of Amsterdam.

de Groot, A.D. 1965. *Thought and Choice in Chess* 2, edited by G.W. Baylor. Mouton Publishers: The Hague-Paris-New York.

Hamburg, F. 2007. Kunnen Kennissystemen Professionals Helpen Medisch-Ethisch Beter te Beslissen? *Liber Amicorum in honour of the Sixtieth Birthday of H. Jaap van den Herik*: 187-199.

van den Herik, H.J. 1991. *Kunnen Computers Rechtspreken.* Inaugural Address, Leiden University.

van den Herik, H.J. 2016. *Intuition is Programmable.* Valedictory Address, Tilburg University.

Herschberg, I.S. 1978. *In de ban van de fout.* Inaugural Address, Delft University of Technology.

Koot, M.R. 2012. *Measuring and Predicting Anonymity.* PhD. thesis, University of Amsterdam.

LeCun, Y., Bengio, Y., and Hinton, G.E. 2015. 'Deep Learning'. *Nature.* 52: 436-444.

Morgan, J. 2014. Privacy is Completely and Utterly Dead, And We Killed It. *Forbes*, August 19.

Royal Netherlands Academy of Arts and Sciences 2016. *Ethical and Legal Aspects of Informatics Research.* Amsterdam, September, KNAW.

Seirawan, Y. 1997. The Kasparov- DEEP BLUE Match' *ICCA Journal.* 19:1, 38-41.

Van de Voort, M., Pieters, W., and Consoli, L. 2015. Refining the ethics of computer-made decisions: a classification of moral prediction by ubiquitous machines. *Ethics and Information Technology.* DOI 10.007/110676-015-9360-2.

Weber, C.R.M. 2017. *Real-time Foresight – Preparedness for Dynamic Innovation Networks.* PhD thesis. Leiden University (forthcoming).

THE NEW IMBROGLIO[1]
LIVING WITH MACHINE ALGORITHMS
Mireille Hildebrandt

Every day a piece of computer code is sent to me by e-mail from a website to which I subscribe called IFTTT. Those letters stand for the phrase 'if this then that,' and the code is in the form of a 'recipe' that has the power to animate it. Recently, for instance, I chose to enable an IFTTT recipe that read, 'if the temperature in my house falls below 45 degrees Fahrenheit, then send me a text message.' It's a simple command that heralds a significant change in how we will be living our lives when much of the material world is connected—like my thermostat—to the Internet.

Sue Halpern[2]

Since the present futures co-determine the future present, predictions basically enlarge the probability space we face; they do not reduce but expand both uncertainty and possibility. The question is about the distribution of the uncertainty and the possibility: who gets how much of what?

Mireille Hildebrandt[3]

IFTTT stands for 'if this than that'.[4] IFTTT is how computers 'think'. It suggests that computers can only run like closed systems that are deterministic by definition. Nevertheless, due to their processing power, connectivity, and our inventiveness, computing systems have now reached unprecedented levels of complexity, generating previously unforeseen levels of indeterminacy. It turns out that a recursive series of deterministic instructions (IFTTT) is capable of producing emergent behaviours that surprise even those who wrote the initial recipes (an algorithm, ultimately, is nothing more or less than a rather precise recipe or set of instructions). This is the result of advances in a sub discipline of artificial intelligence, called 'machine learning' (ML). We should, however, not mistake deterministic computing systems that follow clear and simple rules to provide *automation* of well-defined tasks (IFTTT), for systems that reconfigure their own behaviours to *autonomically* improve their performance (ML). There is flexibility, a recursiveness and an unpredictability in ML that is absent in 'dumb' IFTTTs. The term 'dumb' here is not meant in a pejorative sense; it merely refers to non-learning algorithms that do not adapt their own IFTTTs on the basis of their 'experience', though they may adapt their behaviour, based on their IFTTTs. A simple thermostat runs on a 'dumb' algorithm: its IFTTT determines that whenever the temperature drops below (or above) a certain degree the heating or air conditioner will be turned on – until that same temperature is reached. A smart energy grid will require a continuous learning process, to calibrate energy demand and (local and centralised) energy supply in a way that enables load balancing as well as energy efficiency. As I hope to clarify, 'dumb' can be smart as far as dependence on computing systems is concerned.

In this essay I will suggest that 'dumb' IFTTTs and ML can each have added value as well as drawbacks, depending on how they are used for what purpose. On the one hand, *automated decision systems* in public administration may, for instance, have the added value of being predictable while sustaining accountability, precisely because they are 'dumb' (they do not learn, they just do as instructed). The drawback will probably be that automated decisions are rigid and may be flawed because only a limited set of data points is taken into account. On the other hand, *autonomic decision support systems* for medical diagnosis may, for instance, have the added value of coming up with unforeseen correlations between previously unrelated data points, precisely because such systems improve their performance due to recurrent feedback cycles. Here the drawback may be that the system is not transparent (its inner operations are black boxed) and its output cannot be

---

1 An imbroglio has been defined as 'a confused mass; an intricate or complicated situation (as in a drama or novel); an acutely painful or embarrassing misunderstanding', cf. 'Imbroglio' Merriam-Webster.com, Merriam-Webster, accessed 2 August 2016. In this essay the concept is used as a reference to the complex entanglement of deterministic and learning algorithms that (in)form our increasingly data-driven environment.
2 Halpern 2014. See also *ifttt.com.*
3 Hildebrandt 2016.
4 Halpern 2014.

explained, other than referring to often-irretrievable statistics. I believe that a discussion that turns on whether one is for or against algorithmic governance in general would ignore the difference between two types of algorithms and thereby obfuscate what is at stake. Instead, the discussion about algorithms should focus on the type of problems that benefit from either strict application of 'dumb' algorithms or from adaptive algorithms that are not entirely predictable.

This being said, the question of when to use what algorithms is neither a purely technical question (if there is such a thing) nor a purely political one (if there were such a thing).[5] The decision to engage either 'dumb' IFTTTs or ML may have far reaching consequences for those subjected to the outcome of algorithmic machines, which turns it into a political question. Even if the outcome is checked or applied by a human person, the impact of algorithmic governance is momentous, because that person may not be capable of explaining the outcome and she may have no competence to amend it. If democracy is about self-government we need to find ways and means to involve those affected by algorithmic governance in the choices that must be made.[6] Democratically informed decisions on what algorithms to use will therefore require public understanding of how their employment is constraint by what is possible and feasible in computational terms.[7] This may sound like an insurmountable challenge, but it is not unlike the challenge of alphabetising an entire population, which ultimately enabled self-government since the era of the printing press. On top of that, the political question depends on coming to terms with the *distribution* of beneficial and adverse effects amongst citizens, commercial and governmental players. It may be that a small set of players benefits, whereas others pay the costs. To chart such effects, we need to understand the difference between automation ('dumb' IFTTTs) and autonomics (ML). As to the latter, I will suggest that the distribution of benefits and costs is contingent on the *distribution of the uncertainty* that is created by predictive and pre-emptive analytics. Though one may intuitively assume that predictions reduce uncertainty, this is agent-dependent. Those with access to the predictions have additional reasons to change course, whereas those who remain in the dark are confronted with decisions they could not have foreseen.

Finally, this brief essay will draw the line between, on the one hand, legal certainty and, on the other hand, both arbitrary decision-making and rigid application of inflexible rules. The Rule of Law aims to create an institutional environment that enables us to foresee the legal effect of what we do, while further instituting our agency by stipulating that such effect is contestable in a court of law – also against big players.[8] Such a – procedural – conception of the Rule of Law implies that both automation and autonomics should be constraint in ways that open them up to scrutiny and render their computational judgements liable to being nullified as a result of legal proceedings. This will not solve all of the problems created by algorithmic governance. It should, nevertheless, create the level playing field needed to partake in the construction of the choice architectures that determine both individual freedom and 'the making of' the public good.[9]

Automation and Autonomics in an Onlife World

The idea of an 'onlife' world was initiated by Floridi and taken up by a group of philosophers, social scientists, neuroscientists, and computer scientists who wrote the Onlife Manifesto, on 'being human in a hyperconnected era'.[10] The original idea was to signal the conflation of online and offline, as this distinction is becoming increasingly artificial. In my book on *Smart Technologies and the End(s) of Law*,[11] I have further developed the notion of an onlife world, suggesting that autonomic computing systems develop a specific kind of mindless agency that animates our new 'social' environment. The onlife world is not merely a matter of turning everything online but also a matter of *things* seemingly coming *alive*. This relates to the difference between automation and autonomics.[12]

5 On the relationship between technology, morality, political issues and law: Chapter 7 and 8 in Hildebrandt 2015a.
6 Cf. Hildebrandt and Gutwirth 2007, where we discriminate between aggregate, deliberative and participatory democratic practices. Instead of discussing these practices in terms of either/or we propose that each has an important role to play. On participatory democratic theory see notably Dewey (1927) and the excellent analysis of Marres (2005). Democratic participation does not assume that consensus can be reached, but vouches that those who suffer the consequences of a policy must be involved, cf. Mouffe (2000), who speaks of agonistic debate as a precondition for sound democratic decision making.
7 Cf. Wynne 1995.
8 I agree with Waldron (2008) that the core of the Rule of Law depends on an effective right to see to it that justice is done, by appealing to an independent court that has authority to decide the applicable interpretation of the law. 7 The concept of a 'choice architecture' refers to the type of choices one can and cannot make and the default settings that favour specific options, taking note that most choices are made implicitly, based on heuristics rather than rational deliberation. See Thaler and Sunstein 2008, and – more interesting - Gigerenzer 2000. 9 Hildebrandt 2015a.t
9 The concept of a 'choice architecture' refers to the type of choices one can and cannot make and the default settings that favour specific options, taking note that most choices are made implicitly, based on heuristics rather than rational deliberation. See Thaler and Sunstein 2008, and – more interesting - Gigerenzer 2000.
10 Floridi 2014. I was part of the Onlife Initiative, see ec.europa.eu/digital-single-market/en/onlife-original-outcome.
11 Hildebrandt 2015a.
12 Hildebrandt 2011.

Since the advent of the steam engine and electricity we are familiar with the automation of menial tasks, delegating physical labour to machines, either because of their enhanced 'horse power' or because of their ability to endlessly and rapidly repeat specific tasks.[13] Currently, however, a new type of automation has developed, automating cognitive tasks that require autonomic behaviour.[14] Mere repetition will not do here, as autonomic systems must be capable of reconfiguring their own rules when responding to changes in the environment.[15] Though both automation and autonomics operate on the basis of algorithms, the first are static whereas the second are adaptive, dynamic, and more or less transformative. Though both can be black boxed by whoever employs them,[16] algorithms that generate machine learning are inherently opaque – even for those who develop and employ them. This is evident when the system runs on deep learning multi-level artificial neural networks that conduct unsupervised learning, but even in other cases there is no easy way to explain why the algorithms decided as they did. Machine learning is an inductive process, its output is liable to inductive bias (bias in the data as well as bias in the algorithms) and its usage is liable to various types of inductive fallacies (e.g. mistaking the outcome for the truth, or deriving an 'ought' from an 'is'). This entails that, as with inductive science, we need certain scepticism as to the extrapolation of patterns detected in the observed data (called the training set in ML) to similar patterns in new data (called the test set in ML). Not only because these new data may contain a black swan, but also because we could probably have used other patterns to describe the initial data points and these other patterns may turn out to better fit with the mechanisms that determine both the training and the test set.[17]

In other words, the temporality of our being rules out that either human or machine learning is infallible. ML does not reduce uncertainty but extends it by adding new predictions that will trigger new responses that in turn call for updated predictions.[18] Moreover, a prediction is a 'present future' that influences the 'future present' because actors will change their behaviour based on such a prediction.[19] ML, based on predictive analytics, will therefore create new uncertainties, since we do not know how actors will respond to the new range of 'present futures'.[20] To the extent that the capability to anticipate uncertainty is a crucial characteristic of living organisms, ML can indeed be said to turn our machine environment onlife. We may even come to a point where 'dumb' algorithms will come to the rescue, consolidating and stabilising cascading uncertainties by simply acting as stipulated, behaving as coded, contributing to a reasonable level of predictability.

It is not that simple, of course. If the onlife world is an imbroglio of 'dumb' as well as smart algorithmic governance, the question will be when to endorse either one and how to foresee their interoperability (or, the lack thereof). As to the frustrations generated by automation let me provide a topical example. Imagine entering the Leiden Railway Station during maintenance work on the tourniquets, so it is not possible to check-in with your public transport chip card. Those at work suggest you can just go through without checking in. At the next station, when checking out, you are automatically fined for traveling without having checked-in. When calling the help-desk the lady ensures you that this is no problem because you will get your money back. However, she ends the conversation with a warning: you can only get your money back three times per year – after that you will have to pay. Trying to explain to her that this was not your mistake does not ring any bells with her; she is just repeating the rules.[21] Note that 'dumb' IFTTTs cannot adjust their own rules based on feedback, which also means that any wrong input will cascade through the system (errors are also automated). Smart systems may be more flexible and improve their performance based on recurrent feedback loops. The question remains, however, who determines the performance metric in the light of what goals. Moreover, as indicated above, the opacity of ML systems may reduce both the accountability of their 'owners' and the contestability of their decisions.[22]

The Political, the Technical, and the Legal
The question of when to employ what type of algorithms is both a political and a technical question. At some point it should also be a legal question, because under the Rule of Law individuals should

13 See Latour (2000) on delegation to technologies.
14 Chess, Palmer, and White 2003, Hildebrandt and Rouvroy 2011.
15 Steels 1995.
16 Pasquale 2015.
17 Mitchell 2006, Wolpert 2016.
18 Gabor 1963.
19 Esposito 2011.

20 Hildebrandt 2016.
21 The example modulates a similar experience of colleagues Aernout Schmidt and Gerrit-Jan Zwenne, as recounted during the Annual Meeting of the Netherlands Lawyers Association, 10th June 2016.
22 For attempts to chart the legal issues of automated and autonomic bureaucratic decision making, see Citron 2007 and Citron and Pasquale 2014.

have effective means to challenge decisions that violate their rights and freedoms. Lawyers call this the right to effective remedies to uphold one's fundamental rights (e.g. codified in art. 13 European Convention of Human Rights). Algorithmic governance easily implies that one is not aware of how decisions have been prepared, moulded or even executed in the intestines of various computational systems. Autonomic computing systems, however, enable the profiling, categorising and targeting as citizens or consumers in terms of high or low risk for health, credit, profitable employment, failure to pass a grade in one's educational institution, for tax and social security fraud, for criminal or terrorist inclinations, and in terms of access to buildings, social security, countries or medical assistance. Such personalised targeting will determine what cognitive psychologists and behavioural economists call 'the choice architecture' that decides which options individuals have, and whether and how these options are brought to the attention of the 'user'.[23] It enables subliminal influencing of individual people, based on techniques like AB research-designs that trace and track how we respond to different interfaces, approaches, and options.[24] To the extent that 'dumb' algorithms rely on the input generated by ML the problems they generate are expounded. This results in an imbroglio of invisibly biased decision systems that mediate our access to the world (search engines, online social networks, smart energy grids and the more), potentially creating unprecedented uncertainty about how our machine-led environment will interpret and sanction our behaviours.

Such gloomy prophecies need not, however, come true. We have struggled against the arbitrary rule of dictators as well as the power of private actors capable of twisting our hand. We have developed ways and means to protect human dignity and individual liberty, achieving the kind of legal certainty that safeguards both the predictability and trustworthiness of our social and institutional environment and its open texture in the face of legitimate argumentation.[25] The point is that we cannot take for granted that remedies that worked in the era of printing press, steam engine, and electricity will necessarily protect us in an onlife world. This will require rethinking as well as reinventing the Rule of Law, for instance by making the intestines of the emerging imbroglio transparent and by making its decisions contestable. In recent articles the so-called right to profile transparency of the EU General Data Protection Regulation has been heralded for its spot-on approach to ML algorithms.[26] This right means that automated decisions (whether based on dumb or smart algorithms) that significantly affect people, trigger the fundamental right to data protection. More specifically, such decisions 'automatically' generate two obligations and one right: first, people must be told about the existence of automated decisions, second, they must be given meaningful access to the logic of such decision, and, third, those concerned have a right to object against being subject to such decisions. It is interesting to note that the right to profile transparency is framed – by some - as a clash between US based AI companies and the EU, or even between innovation and Luddite hesitation.[27] I believe that such labels are out-dated and stand in the way of global progress. Progress involves hesitation as well as innovation, high risk and high gain, but not at the cost of those already disadvantaged, or vulnerable to data-driven exclusion. At some point any individual person faced with the onlife imbroglio - that we are already a part of – may be disadvantaged by and vulnerable to unfair and degrading treatment by interacting automated and autonomic computing systems.

Profile transparency implies that the uncertainty generated by ML should be contained, to prevent mishaps. It also implies that decisions that seriously affect individuals' capabilities must be constructed in ways that are comprehensible as well as contestable.[28] If that is not possible, or, *as long as* this is not possible, such decisions are unlawful. In that case we may have to employ dumb algorithms, though even the outcome of dumb algorithms must be comprehensible and contestable. As to ML, we need to invest in the engineering of choice architectures that re-instates our agency instead of manipulating it. This is not about 'the more choice the better'.[29] It can only be about involving those whose onlife is at stake in the construction of the choice architectures that will define their capabilities – and thus, their effective rights.

23 Thaler, Sunstein and Balz 2010.
24 This also regards attempts to influence voting, e.g. Christian and Griffiths 2016, explaining AB research-design and its usage in the US presidential elections. On the capacity to influence voting by merely tweaking a search algorithm see Epstein and Robertson 2015.
25 Waldron 2011. See also, arguing for legality and against legalism

Hildebrandt 2015b.
26 E.g. Goodman and Flaxman 2016, cf. Hildebrandt 2012.
27 E.g. Metz 2016.
28 My use of the term capability in inspired by Sen (2004), where the capability is the substance that is to be protected, while the right itself co-constitutes the capability by safeguarding its sustainability.
29 Van den Berg 2016.

References

Van den Berg, B. 2016. Coping with Information Underload: Hemming in *Freedom of Information through Decision Support. In Information, Freedom and Property:* The Philosophy of Law Meets the Philosophy of Technology, edited by Mireille Hildebrandt and Bibi van den Berg. Abingdon, Oxon UK: Routledge.

Chess, D. M., Palmer, C. C., and White, S. R. 2003. Security in an Autonomic Computing Environment. *IBM Systems Journal* 42:1, 107–18.

Christian, B. and Griffiths, T. 2016. *Algorithms to Live By: The Computer Science of Human Decisions.* New York: Henry Holt and Co.

Citron, D. K. 2007. Technological Due Process. Washington University Law Review 85, 1249–1313.

Citron, D. K. and Pasquale, F. 2014. The Scored Society: Due Process for Automated Predictions. Washington Law Review 89:1, 1–33.

Dewey, J. 1927. *The Public & Its Problems.* Chicago: The Swallow Press.

Epstein, R. and Robertson, R. E. 2015. The Search Engine Manipulation Effect (SEME) and Its Possible Impact on the Outcomes of Elections. *Proceedings of the National Academy of Sciences* 112:33, E4512–21.

Esposito, E. 2011. *The Future of Futures: The Time of Money in Financing and Society.* Cheltenham: Edward Elgar Publishing.

Floridi, L. 2014. *The Onlife Manifesto - Being Human in a Hyperconnected Era.* Dordrecht: Springer.

Gabor, D. 1963. *Inventing the Future.* Secker & Warburg.

Gigerenzer, G. 2000. *Adaptive Thinking: Rationality in the Real World.* Oxford; New York: Oxford University Press.

Goodman, B. and Flaxman, S. 2016. European Union Regulations on Algorithmic Decision-Making and A "right to Explanation". *arXiv:1606.08813 [Cs, Stat]*, June. arxiv.org/abs/1606.08813

Halpern, S. 2014. The Creepy New Wave of the Internet. *The New York Review of Books.* www.nybooks.com/articles/2014/11/20/creepy-new-wave-internet/

Hildebrandt, M. 2011. Autonomic and Autonomous "Thinking": Preconditions for Criminal Accountability. In *Law, Human Agency and Autonomic Computing.* Abingdon: Routledge.

———. 2012. The Dawn of a Critical Transparency Right for the Profiling Era. In *Digital Enlightenment Yearbook 2012.* Amsterdam: IOS Press. 41–56.

———. 2015a. *Smart Technologies and the End(s) of Law. Novel Entanglements of Law and Technology.* Cheltenham: Edward Elgar.

———. 2015b. Radbruchs Rechtsstaat and Schmitt's Legal Order: Legalism, Legality, and the Institution of Law. *Critical Analysis of Law* 2:1. cal.library.utoronto.ca/index.php/cal/article/view/22514

———. 2016. New Animism in Policing: Re-Animating the Rule of Law? In *The SAGE Handbook of Global Policing,* edited by B. Bradford, B. Jauregui, I. Loader, and J. Steinberg. London: SAGE.

Hildebrandt, M. and Gutwirth, S. 2007. (Re)presentation, pTA Citizens' Juries and the Jury Trial. *Utrecht Law Review.* Accessed at *www.utrechtlawreview.org/* 3, 1.

Hildebrandt, M, and Rouvroy, A. 2011. *Law, Human Agency and Autonomic Computing. The Philosophy of Law Meets the Philosophy of Technology.* Abingdon: Routledge.

Latour, B. 2000. The Berlin Key or How to Do Words with Things. In *Matter, Materiality and Modern Culture,* edited by P.M. Graves-Brown. London: Routledge. 10–21.

Marres, N. 2005. *No Issue, No Public. Democratic Deficits after the Displacement of Politics.* Amsterdam, available via: http://dare.uva.nl.

Metz, C. 2016. Artificial Intelligence Is Setting Up the Internet for a Huge Clash With Europe. *WIRED,* July 11.

Mitchell, T. M. 2006. *Machine Learning.* Mcgraw-Hill Computer Science Series

Mouffe, C. 2000. Deliberative Democracy or Agonistic Pluralism. Department of Political Science, Institute for Advanced Studies, Vienna.

Pasquale, Frank. 2015. *The Black Box Society: The Secret Algorithms That Control Money and Information.* Cambridge: Harvard University Press.

Sen, Amartya. 2004. Elements of a Theory of Human Rights. *Philosophy & Public Affairs* 32: 4, 315–56.

Steels, Luc. 1995. When Are Robots Intelligent Autonomous Agents? *Robotics and Autonomous Systems.* 15:

3–9.

Thaler, Richard H., and Cass R. Sunstein. 2008. *Nudge: Improving Decisions about Health, Wealth, and Happiness.* New Haven: Yale University Press.

Thaler, Richard H., Cass R. Sunstein, and John P. Balz. 2010. Choice Architecture. SSRN Scholarly Paper ID 1583509. Rochester, NY: Social Science Research Network. http://papers.ssrn.com/abstract=1583509.

Waldron, Jeremy. 2008. The Concept and the Rule of Law. *Georgia Law Review* 43:1, 1.

———. 2011. Are Sovereigns Entitled to the Benefit of the International Rule of Law? *European Journal of International Law* 22:2, 315–43.

David H. 2013. Ubiquity Symposium: Evolutionary Computation and the Processes of Life: What the No Free Lunch Theorems Really Mean: How to Improve Search Algorithms. *Ubiquity* 2013 (December): 2:1, 2-15.

Wynne, Brian. 1995. Public Understanding of Science. In *Handbook of Science and Technology Studies*, edited by Sheila Jasanoff, Gerald E. Markle, James C. Petersen, and Trevor Pinch. Thousand Oaks, CA: Sage. 361–89.

060

FREEDOM AND DATA PROFILING
Liisa Janssens[1]

Algorithms normally behave as they're designed, quietly trading stocks or, in the case of Amazon, pricing books according to supply and demand. But left unsupervised, algorithms can and will do strange things. As we put more and more of our world under control of algorithms, we can lose track of who – or what – is pulling the strings.

Christopher Steiner[2]

What would happen if a set of algorithms could identify our behaviour and possibly even our thoughts? Such sets of algorithms would drastically change our interactions and relationships as human beings. Today, sets of algorithms can already generate profiles on the basis of data about an individual user. In doing so, an 'average you' is mapped; this process is called data profiling. In data profiling, however, lies the risk that human beings will only be seen through conceptions that follow from a system that is guided by probability analyses. At a time when data analyses are increasingly present in our day-to-day reality, we ought to reflect on the question as to whether human beings can be completely categorised on the basis of their data profile, or whether man as 'the Other' also contains a mystery that reaches further than these analyses can reach.

The previous question is twofold. The first part concerns the consequences of the origin, We should, the area of the underlying technology of algorithms, wonder who or what the makers of these algorithms are, to what extent errors can occur in the process of programming, and whether we are able to detect such errors. The second part concerns the method, when analysing big data sets, we should ask ourselves what kind of methodology is used in combining these data sets. In order to illustrate the urgency of the foregoing problems, I will elaborately discuss the implementation of The Internet of Things, a development that will bring increasingly more data streams into our daily lives and that will usher in 'a Hyperconnected Era'.[3]

Secondly, in the light of 'a Hyperconnected Era', I will question from a meta-ethical point of view the influence of algorithms on our freedom to see 'the Other': a non-categorised human being. The philosophers Jacques Ellul and Nicolai Berdyaev approach freedom at a meta-ethical level. These thinkers do not deny that man is bound by empirical reality and should therefore relate to systems, but they also proceed from meta-ethical principles that are interpersonal: it is in these principles that freedom lies, according to Ellul and Berdyaev, as they are transcendent and not bound by any system. In this approach, ethical values are created, in freedom, from one person to 'the Other' *as a person*.

Can the notion of freedom that is central in the philosophy of Ellul and Berdyaev provide a counterweight to the strict and deterministic way of thinking, which objectifies and reduces human beings to packages of data that underlies the practice of big data profiling? To what extent does the freedom to be in a relationship with 'the Other' as *another person*, outside of the system of profiling, by cherishing hope and taking personal responsibility, remain to mankind?

Increase in Data Streams Will Usher in 'a Hyperconnected Era'

When thinking about the meta-ethical implications of technological devices, which generate data streams, it is important to approach the issue from the point of view of the set of devices. It is for this reason that I will not discuss the developments surrounding drones, smartphones, or tablets separately. But I will rather discuss the whole system of these devices, which will generate an almost incomprehensible amount of data streams and exchange this data within the system; this development will usher in 'a Hyperconnected Era'. Large technology firms are collectively expanding their arsenal of means that enable the collection of data. They are so successful in doing so, that the question of 'why' or 'wherefore' is rarely asked. This is probably due, partly, to the level of difficulty of the ethical questions as to the possibilities that sets of algorithms provide for analys-

1 My sincere thanks to Marleen Postma, for her friendship and support during the process of shaping this text.
2 Steiner 2012: 5.
3 Floridi 2016: The notion of 'a Hyperconnected Era' is taken from The Onlife Manifesto.

ing data. The fundamental rights and freedoms that normally provide some degree of support, such as privacy provisions in laws and regulations, provide little solace when ethical questions regarding big data are concerned. Even though the will to keep up with every new development is there, it sometimes seems like a Sisyphusean task to do so. One problem is that the flow and collection of data do not stop at the borders of states and their jurisdictions, causing the unwieldy system of laws and regulations to always follow the facts. However, the process of expanding big data analyses through technological means did not suddenly invade our lives; it entered step by step. Due to the difficulty of the new ethical questions that arise in response to data profiling in 'a Hyperconnected Era', it is no surprise that big data analyses enter our lives without any significant resistance. Steiner states that:

> It may sound like a far-off prospect, but algorithms have already figured many of us out, parsing our personalities and learning our motivations. We're rarely given explicit warning when a bot is observing us; it's something that's crept into our lives with little fanfare.[4]

In 2014, Facebook bought messaging service Whatsapp for 22 billion US dollars. World wide, the messaging service had approximately 450 million active users that year. This acquisition illustrates the exorbitant value of data companies; acquisitions happen primarily for profit, from the point of view of both the buying and the selling party. Acquisitions generally do not happen in order to serve individual users or to protect their interests, even though that is the marketing strategy that is initially employed for attracting users. Even today, the legal information starts with the line: 'Respect for your privacy is coded into our DNA.'[5] WhatsApp was initially advertisement free, but rather than offering an alternative to Facebook, the business chose instead to monetise its data. Soon, this data will be shared with parent company Facebook, in order to enable the sending of targeted ads, adjusted to an 'average you'. Avoiding this proves troublesome. You can disable this option, but when your WhatsApp contacts have agreed to targeted advertising and receive individual messages about their travel plans to their holiday destinations, is your data, since you are a contact in their address book, not automatically included as well?[6]

Even though WhatsApp users expressed privacy concerns during the take-over in 2014, resulting in a short-lived boycott, the messaging service has nearly a billion users in 2016.[7] Google's presence, too, continues to expand. The company by now has an extensive reach: '–into the home (Nest), car (Waze), space (satellite investments), and workplace (Google Enterprise), and its ability to buy data from hundreds of brokers, makes 'total information awareness' by the company less a paranoid fear than a prosaic business plan.'[8] Algorithms can potentially be used in a valuable way, as they continue to discover more about us. The Internet of Things, which exists by virtue of data streams and algorithms, is expected to generate 1.9 trillion US dollars of Economic Value Added (EVA).[9] Sets of algorithms can indeed improve our lives in many domains, for example in health care, when, in prescribing medications, doctors can take into account an 'average you' rather than an 'average human being' on the basis of your medical file and those of other patients. The knowledge that data analyses can provide might enable us to live longer, or simply provide us with more comfort. However, this is only one side of the coin. After all, it is an illusion to think that mankind will only gain more freedom through the possibilities that big data analyses offer. Even though these analyses give us the impression that we have a great (technological) power over our lives, our freedom may, paradoxically, have decreased. Algorithms have been programmed and during the programming process, (ethical) choices have been made and mistakes could have occurred. The danger is that people start thinking in terms of the functionality of the system of analyses, but that they no longer question the technology itself.

Will the user get insight into his or her data profile? Will there be a possibility for the person who is the subject of the data profile to correct any mistakes in this profile, if they want to? Do the programmers even know which ethical choices they have made and what the consequences of these choices are for users? One of the important developments that will increase the number of

4 Steiner 2012: 164.
5 WhatsApp 2016.
6 WhatsApp 2016.
7 Newspaper Trouw 2016: 1.
8 Pasquale 2015: 189.
9 Gartner 2013.

analyses of the 'average you' on the basis of data streams is the implementation of The Internet of Things (examples of this are Google's Nest and Waze, mentioned above). This implementation brings about a transition in which the digital will increasingly merge with the material world. Taking into account the fact that in 2009, 2,5 billion things had access to the Internet already – a number that we expect to grow to 30 billion by 2020 –, mankind seems unable to evade a future of hyperconnectivity.[10] The largely invisible hyperconnectivity of data flows and the probability analyses that are associated with this technology could change our conceptions and our encounter with the other as a fellow human being. Generated data analyses about a human being, for example assumptions about one's behaviour, rely upon calculations of probability. One of the characterstics of big data is that the accuracy of the data is not always guaranteed. Big data analyses suffer from methodological problems, as well-founded scientific research is still lacking; after all, many applications have been in use for too short a period of time to be properly evaluated. The available studies were often carried out poorly and the results are incomplete.[11] These shortcomings make it difficult to assess whether the use of big data analyses is effective or even justified.[12] In big data processes, the responsibility for the accuracy of the data is under pressure. In the majority of cases, this responsibility lies with the organisation that collects and/or owns the data. In some cases, this responsibility is extended to the organisation that analyses the data. When organisations share, edit, and re-contextualise data, there is no longer one party that clearly bears responsibility for the quality and accuracy of the process and the end result.[13] Problems of responsibility constitute an issue that cannot be avoided and that becomes increasingly more important as big data processes continue to develop and be implemented into our daily lives. The same problems emerge when multiple parties work with and have access to a database of which the data is confidential and ought to be stored safely.[14] Underneath the problems described above lie meta-ethical questions concerning responsibility. When one claims that big data analyses have a value of probability over empirical reality, the problem of social determinism reveals itself: human beings are only being seen through his or her data profile. When big data analyses receive the label that these contain information about reality, or maybe even a truth about us or 'the Other', it is questionable whether, as these analyses get increasingly common, decisions or actions that concern human beings can be taken on the basis of the outcomes of these analyses. Can we delegate responsibility for 'the Other' to a system consisting of sets of algorithms, or should we, as human decision makers outside of the system, take personal responsibility? Will a system consisting of sets of algorithms determine the sentencing of a defendant on the basis of big data analyses of the risk of reoffending? Or is a human decision maker necessary, as criminal law constitutes an *ultimum remedium* that has a deep impact on the life and liberty of man? To what extent do sets of algorithms provide reliable information about our reality and to what extent should we adjust our actions to probability analyses?

Big Data Profiling As a Dystopia

Social determinism through data profiling implies that man is approached as no more than a biological being. Via a system of sets of algorithms, big data analyses provide a conception of man in reality, but to what extent is there, and can there be, room for a truth outside of the system? In light of the possibilities that big data analyses offer, we can draw an interesting analogy with the science fiction film *Gattaca*,[15] in which the danger of social determinism on the basis of DNA analyses is portrayed. In *Gattaca*, the fate of man as a biological being is determined through a system of DNA profiling. On the basis of their DNA profiles, decisions regarding the social position of human beings, not as free individuals but as predetermined beings, are justified. In *Gattaca*, protagonist Vincent Freeman (Ethan Hawke) dares to challenge the system of profiling. He was born the old-fashioned way, without embryonic preselection, and when he was born, a probability analysis of his DNA predicted that he would have heart problems, poor vision, and a life expectancy of approximately 30 years. It is for this reason that he was socially predetermined as a so-called 'In-Valid', a classification that places him at the bottom of the social ladder; because of his genetic profile, he is forced to work as a cleaner. In spite of his social status, he dreams of

10 Gartner 2013.
11 WRR 2016: 111.
12 WRR 2016: 109.
13 WRR 2011.
14 WRR 2016: 111.
15 *Gattaca*. Directed by Andrew Niccol, Oct. 24, 1997, Colombia Pictures.

becoming an astronaut with the *Gattaca* Aerospace Corporation and going to one of the moons of Saturn on a mission. This future, however, is only reserved for those with the best genes, for people like his brother Anton Freeman (Loren Dean). In spite of his low social status, Vincent cherishes hope and with indefatigable determination, he goes against the system. In spite of his status of 'In-Valid', he beats his 'Valid' brother multiple times during high risk swimming competitions in open sea. In fact, he even has to save his brother from drowning multiple times. With the help of a gene broker, Vincent comes into contact with Jerome Eugene Morrow (Jude Law), who has the perfect genes, but was disabled by an accident. They exchange DNA profiles and with the dedicated help of others who also do not conform to the system, Vincent is given the chance to not give up hope and make his dream come true.

Even though (DNA) profiling as it is shown in the film *Gattaca* is not yet one of our possibilities, big data analyses nowadays do provide us with the possibility of generating individual data profiles. These profiles are not profiles of the average individual from a certain area or a certain age category, as would generally be the case with statistical analyses; instead, they are profiles of the 'average you'. The possibilities that big data analyses offer to generate this kind of specialised profile of an individual raise ethical questions regarding social determinism through data profiling. With the possibilities offered by big data analyses, the social determinism we see in *Gattaca* is drawing ever closer. After all, when big data analyses take on an increasingly prominent role in our conceptions of reality, this could have consequences for our freedom as human beings to go against the system of data profiling. It is perfectly conceivable that a change in our conceptions of the other as a fellow human being could influence our interpersonal values and social positions.

Man As a Predetermined Being?

The technological developments described above, which can provide a probability analysis of for example someone's medical condition or even predict someone's thoughts and (future) behaviour, create the impression that the world of mystery is slowly fading away. The DNA profiling that we see in the film *Gattaca* could be a vision of the future that is closer than we think. An important remark that needs to be made in this regard is that data analyses can also be the result of a combination of several, methodologically not matching, sets of data, causing the results of the analyses to not have any valid value of probability over this reality. However, even when the data analyses are conducted in a methodologically sound manner, they are no more than calculations of probability, yet there is a danger that people will be inclined to shift responsibility onto these analyses. The question is whether man can only be categorised as a biological being or whether inside man, there is a mystery that cannot be grasped by a system. The vision of Jacques Ellul is that even though man is a determined being, it is up to man to transcend this necessity. It is in this act that man's freedom lies.[16] An application of this vision to 'a Hyperconnected Era' involves an awareness of the necessity to provide a counterweight to this new form of determinism that is made possible by algorithms and the system of data analyses and data profiling. The question is not to be saved from this new form of determinism; we ought to relate to it in empirical reality. However, each individual can challenge this system of determinism in an act of freedom in order to bring about personal change. Ellul's philosophy is an appeal to personal responsibility and it is in this regard that Ellul calls himself a dialectician. He describes society as totalitarian, but leaves open the possibility of change via personal transformation.[17] What is at stake is the awareness that technology influences our conceptions of man; these conceptions, when we see human beings only via data analyses, reduce 'the Other' to a lifeless object. Man becomes a person when he is aware of the necessity to take measures against this process.[18] The motor of the dialectic, according to Ellul, is not located in empirical reality but in the 'entirely Other'.[19] It is crucial to enter into a dialogue with the person of 'the Other'[20] – which cannot be (pre)determined by whatever system. We also find this approach in the existentialist personalism of Nicolai Berdyaev. He pursues a concrete ethics in which man himself is the creator of values. What is at stake in his philosophy is not a normative ethics that can be rationalised and incorporated into, for example, a system

16 Achterhuis 1992: 48.
17 Achterhuis 1992: 48.
18 Kristensen 1986: 110.

19 Achterhuis 1992: 48.
20 Kristensen 1986: 32.
21 Harold.

of laws and regulations.[21] In existentialist personalism, freedom is central. Reality is approached as a 'possibility', in which persons can flourish. Berdyaev's philosophy centres on an ethics that is created in freedom from one person to another. This is in stark contrast to the deterministic vision that results from the approach, which holds that reality only receives its validity via the rational, using means such as analyses of probability. Berdyaev emphasises the irreducible value of man. Part of being human is 'to-be-there-in-relation-to-the-Other', outside of a form of knowing that goes via the (pre)determined system.[22] This philosophy proceeds from the dialectical relation that man has with truth in empirical reality. Part of man remains a mystery and man's freedom lies in the possibility to see the other as a person, as 'the Other', outside of the system. The contradiction between the possibilities that big data analyses offer to generate probability analyses of man on the one hand, and the necessity to take responsibility for 'the Other' outside of the system on the other hand, raises the question: What part of man can be a categorised being by knowing and what part of man remains a mystery that is beyond the grasp of human reason?

> However, if we do discover a complete theory, it should in time be understandable in broad principle by everyone, not just a few scientists. Then we shall all, philosophers, scientists, and just ordinary people, be able to take part in the discussion of the question of why it is that we and the universe exist. If we find the answer to that, it would be the ultimate triumph of human reason – for then we would know the mind of God.[23]

Until that time has come, these insights point us, as humanity, to the necessity of continuously asking ourselves the meta-ethical question of how we should re-evaluate our freedom in relation to 'the Other' in 'a Hyperconnected Era'.

065

22 Harold.
23 Hawkin 1988: 94.

References

Gattaca. Directed by Andrew Niccol, Oct. 24, 1997, Colombia Pictures.

Achterhuis, H. 1992. *De maat van de techniek*. Baarn: ambo.

Floridi, L. 2014. *The Onlife Manifesto - Being Human in a Hyperconnected Era*. Dordrecht: Springer.

Gartner, N. 2013. *Gartner Says Personal Worlds and the Internet of Everything Are Colliding to Create New Markets*. Accessed 19 September 2016 at *www.gartner.com/newsroom/id/2621015*.

Harold, P. J. *The Desire for Social Unity: Existentialism and Postmodernism*. Robert Morris University.

Hawkin, S. 1988. *A Brief History of Time - From Big Bang to Black Holes*. Bantam Dell Publishing Group.

Kristensen, B. 1986. *Het verraad van de techniek*. Amsterdam: VU Uitgeverij.

Pasquale, F. 2015. *The Black Box Society*. Harvard University Press.

Scientific Council for Government Policy. 2011. *WRR rapport De staat van informatie*. Amsterdam University Press.

Scientific Council for Government Policy. 2016. *WRR rapport Big Data in een vrije en veilige samenleving*. Amsterdam University Press.

Steiner, C. 2012. *Automate This, How Algorithms Came To Rule Our World*. Portfolio Penguin.

van Teeffelen, K. 2016. Whatsapp deelt toch informatie met Facebook. *Trouw*. 25 August. Accessed at *www.trouw.nl/tr/nl/5133/Media-technologie/article/detail/4364424/2016/08/25/Whatsapp-deelt-toch-informatie-met-Facebook.dhtml*.

WhatsApp. 2016. Accessed 19 September at *www.whatsapp.com/legal/*.

A UTOPIAN BELIEF IN BIG DATA

Esther Keymolen[1]

Today's government agencies increasingly turn to data-driven strategies to steer and regulate society. Looking at some headlines of recent press releases, the first successes seem to be already there; through predictive policing, the Californian police force has brought down robberies by a quarter,[2] the Dutch tax office works with a Risk Indication System to detect tax fraud;[3] and the collection and analysis of large amounts of medical data has helped to personalise medical treatment.[4] These are just a few examples of data-driven policy interventions that are based on what has been called Big Data. Although there is no common definition of Big Data, it generally refers to the automated analysis of – often-combined – data sets, to look for patterns or correlations based on which predictions can be made. Big Data has been widely described as a revolution, which will profoundly change the way we do business, develop policy, and conduct research. Big data should enable governments – and other actors as well – to 'anticipate future needs and concerns, plan strategically, avoid loss, and manage risk.'[5] All in all, Big Data has been discovered as the new 'holy grail' to develop effective, data-driven, and citizen-centred policies.

Although Big Data analysis has its limits and in the end is about probability rather than about certainty, these important nuances are often lost and an all-encompassing, utopian belief in what Big Data has to offer ousts a more critical stance. The idea of Big Data is temping; it promises to make our future no longer a sea of uncertainties in which we may drown, but a safe pool of tangible possibilities in which we can swim. Yet, this temptation tends to make people ignorant for the new complexities that Big Data inevitably also brings forth. This essay explores the philosophical grounds for this utopian belief in Big Data. Where does this deeply felt urge to ignore the transforming workings of technology come from? Based on insights deriving from the philosophical anthropology of German philosopher Helmuth Plessner (1892-1985) and the philosophy of technology, this essay will investigate the unbreakable alliance between human beings and artefacts. Human beings need technology to shape their lives. This shaping, however, is a continuous endeavour. Technology has often unforeseen – and sometimes unwanted – consequences, urging human beings to re-evaluate their inventions. While technologies, such as Big Data, definitely can help to make the uncertainty that a contingent world invokes bearable, human beings, nevertheless, long for more. Humans do not merely want to cope with a contingent world, they want to overcome contingency all together! This essay will show that this utopian longing to solve the problem of contingency is not only unfeasible due to the weight of the technology, but is also undesirable, as it inhibits a denial of the eccentric, playful identity of human beings.

067

A Complex World

All living things live in a complex environment, meaning that their environment contains more possibilities than they themselves can actually realise. As a result, in order to persevere some of this complexity has to be reduced. For human beings, this reduction of complexity is extra challenging as they are consciously aware of the world's *contingency*. Human beings realise that their lives could have been different; that the world they inhabit could have been different. Consequently, human beings know it is up to them to make decisions about how their environment comes to look like. At the same time, they realise that they cannot completely oversee the consequences of their choices.[6] Complexity enters the human world by means of two elements: *the other* and *the future*, revealing a social and a temporal level in the complexity of the world.[7] On the one hand, human beings face the unanticipated actions of others who are, to a certain extent, free to see and do things differently. On the other hand, here and now, human beings have to deal with the idea of an unknown future. Of all the possible futures that might unfold, only one can become reality. The question of course is: which one will it be?

1 I like to thank my colleague Jenneke Evers MA for reading the draft version and providing me with very useful feedback.
2 See privacysos.org/predictive, accessed: 19 July 2016.
3 See www.nltimes.nl/2014/10/02/big-data-key-dutch-fight-tax-fraud, accessed: 19 July 2016.
4 Cf. www.modernhealthcare.com/article/20150822/MAGA-ZINE/308229976, accessed: 19 July 2016.
5 Kerr and Earle 2013: 66.
6 Luhmann 1979; Bednarz 1984.
7 Luhmann 1979.

The reason for this radical complexity is closely connected to the building plan of human beings, their so-called ontology. Human beings do not completely coincide with themselves. Human beings have, what Helmuth Plessner has referred to as, an eccentric positionality.[8] While it is true that human beings experience their actions and thoughts as direct, as an inherent part of themselves, the life of human beings is simultaneously also placed outside of themselves, eccentric. This external position enables awareness and reflexivity. From a distance as it were, they perceive their relation towards themselves, others, and the world around them. It is their eccentric positionality that makes them receptive for the unpredictable behaviour of others and for an over-complex future. It is the distance they experience towards others and their environment that lies at the heart of the complexity that needs to be reduced.

Artificial by Nature

A very effective way of reducing this complexity is through technology. Human beings require technology – in the broadest sense of the word – to build themselves a home in a world of which they sense its contingent character. Plessner argues that 'because of their eccentric positionality, human beings are *homeless by nature*'.[9] It is only through technology and through their creative powers, by repeatedly reinterpreting and carving out their lives, the world human beings inhabit can become a familiar, less complex world.[10] Plessner characterises human beings, therefore, as 'artificial by nature.'[11] They cannot be understood without taking into account the tools they use to shape their lives.[12]

It is conspicuously clear that Big Data currently is one of the preferred tools to reduce complexity and construct a familiar world. The predictive power of Big Data unravels the future and enables governments to anticipate – and steer! – the behaviour of their citizens. In other words, the two main sources of complexity, an unknown future and the behaviour of others, are directly tackled by Big Data. Real-time traffic control, for example, makes it possible to predict where a traffic jam will occur and subsequently enables traffic controllers to redirect drivers in order to prevent congestions.

Technology: a Temporary Shelter

Although technology, or culture, to be more precise, provides human beings with a familiar world, it has to be noted that this can only result in a *temporary shelter*. The furnished world of culture functions as a filter between human beings and the open, world they live in, enabling them to develop a fragile and frequently disturbed balance that lies at the root of their daily life.

On the one hand, this temporary character is caused by the technology itself. Although made by human beings, artefacts gain their own momentum; they have a kind of heaviness that stands apart from the people who created them. Artefacts bring forth unexpected – not necessarily always unwanted – consequences, which challenges human beings to rethink their invention. With every technological intervention, problems are solved and new ones are created. Moreover, artefacts are 'multi-stable.'[13] This means that an artefact can bear different meanings. It is typical for human beings to use artefacts in different contexts, shaping their environment over and over again. The openness of artefacts can be negatively framed as a burden, causing instability in human life. However, it can just as well be defined as a chance, instigating new and fruitful practices or, in other words, prompting innovation.[14]

On the other hand, the temporary character of this equilibrium is brought forth by the eccentric position of human beings themselves. Their ability to take this external position makes that individuals can never be fully captured. They escape a conclusive interpretation. Through technology, they are continuously interpreting the world around them and with this detour they interpret themselves.[15] Technologies mediate and are constitutive for human actions and interactions.[16] All in all, we can say that technology is an inherent part of the lives of human beings, but simultaneously it is not fully under their control.

8 Plessner 1975.
9 Plessner 1975: 291.
10 Keymolen 2016.
11 Plessner 1975: 310.
12 Cf. de Mul 2003: 254.

13 Ihde 1990.
14 Keymolen 2016.
15 Cf. Hildebrandt 2015: 183.
16 Verbeek 2011; 2010.

Several authors have argued that on different levels, this mediation of human beings and technology, also applies to Big Data. For example, Kitchin argues that although we may be able to use algorithms we do not firmly understand their workings and that in fact they challenge the established epistemologies across the sciences.[17] Richards and King depict three paradoxes of Big Data that illustrate how data-driven decisions enable people to interpret and control the environment and at the same time pre-sort society in unforeseen – and often unwanted – ways. First, on the one hand Big Data makes society more transparent, on the other Big Data analysis often operate in the background, invisible for citizens. Second, Big Data promises to provide new insights on who we are and what we will become, but simultaneously the identifying power of Big Data can restrain the dynamic, eccentric character of human identity. And finally, Big Data empowers users because it makes the future more predictable, it however also may cause a power imbalance, as Big Data is not equally accessible to everyone.[18]

Utopian Beliefs

By continuously shaping their environment through technology, the life of human beings is always *under construction*. As artefacts reduce complexity by giving users some control over their environment, all these handy tools, from apps on smartphones to smart algorithms mining databases, simultaneously also introduce new complexity, calling upon human beings to come up with new solutions.

Notwithstanding this constant mediation of human beings and technology, human beings long for simplicity.[19] Plessner speaks of a *utopian standpoint* to pinpoint how human beings, withstanding the eccentric positionality of their existence, strive for stability and certainty.[20] They have the urge to live as if the world is a well-ordered place and to treat social constructs as if they are governed by natural laws. Instead of the perpetual shaping of their lives, they desire – once and for all – to make whole what is shattered by nature. Nevertheless, this utopia is sometimes shaken to its foundations because of the own weight of technology and of the eccentric positionality of human beings, ever escaping, or missing out on, a final ground.

Whereas Plessner perceives that this longing for a final ground leads human beings to the domain of religion, it can be argued that especially in late-modern Western society technology has occupied the place of God.[21] With every new technological invention, there is the promise to overcome the distance human beings experience towards the world around them, others, and towards themselves.[22] Especially nowadays, smooth-operating and pro-active smart and networked technologies, invisibly functioning in the background, feed the longing of people to live in an environment enabled by smart technologies without the interference of those technologies. Ihde speaks revealingly about a 'double desire' of human beings, which:

> On one side, is a wish for total transparency, total embodiment, for the technology to truly 'become me.' Were this possible, it would be equivalent to there being no technology, for total transparency would be my body and senses; I desire the face-to-face that I would experience without the technology. But that is only one side of the desire. The other side is the desire to have the power, the transformation that the technology makes available.[23]

In other words, we want the outcome of technology, without acknowledging the way in which technology also exerts its own weight. We do not just want to reduce complexity; we want to eliminate it.

This one-sided, utopian belief in the power of technology, especially when it comes to Big Data, is conspicuously present in the policy domain. The collecting and mining of data on a large scale has set in motion a new way of approaching reality. The basic belief is that if enough data can be gathered and correlations can be found, it will become possible to predict the future almost completely. Datafication, the process of capturing actions and behaviours into readable

069

17 Kitchin 2014a; 2014b.
18 Richards and King 2013.
19 Cf. Verbeek 2011.
20 Plessner 1975.
21 de Mul 2003.
22 Cf. Aydin and Verbeek 2015: 292.
23 Ihde 1993: 75.

data, holds the promise that society and human behaviour become increasingly quantifiable and therefore predictable, malleable even.[24] The contingency of the world no longer is a mere fact of life, but an 'epistemic failure' which can and must be overcome.[25]

Scholars in Big Data have pointed out that a 100% solid prediction – which then actually is no longer a prediction but a fact – is unfeasible. Not only is Big Data in essence about probability, about assessing chances, the eccentric positionality of human beings also does not let them be captured in all-defining interpretations. While for many technological applications this underpinning statistical approach is sufficient to perform in a desirable way, a satisfying output should not be mistaken for certainty or truth. However, in general, human beings are very poor in assessing chances and probabilities.[26] What in fact is only likely will quickly become sure and proven, especially when the solution comes rolling out of a computer, lacking a clear and understandable explanation of how the result was reached.[27]

The aura of objectivity that surrounds data-driven policy making, should be critically assessed. Numbers do not speak for themselves; there is always an interpretation of data, and this interpretation brings along limits and biases. If we do not explicate these limits and biases and treat data-driven policies as if they are completely objective and true, we run the risk of misinterpretation.[28] All in all, utopian expectations of Big Data are in conflict with the workings of the technology and with the eccentric nature of human life, since the latter does not allow itself to be pinned down that easily.

Beautiful Complexity

Coming to the end of this essay, I would like to add that this utopian desire to finally recover a final ground, to find the certainty human beings lack by default through the use of Big Data, is not only not feasible, it is also not as desirable as it may seem at first sight. A society grounded on too strong a focus on predictability and control hinders the 'societal intelligence and resilience' inherent in human life. In this kind of society 'relations create no surprise.'[29] Although Big Data may help policy-makers to provide safety, we should always keep an eye for the unintentional outcomes of technology or as Nissenbaum puts it:

 070

> In a world that is complex and rich, the price of safety and certainty is limitation. Online as off, [...] the cost of surety – certainty and security – is freedom and wide-ranging opportunity.[30]

Generally, the lives we strive for are filled with beautiful, yet complex things, such as art, music, love, and friendship. A life stripped from its complexity is a boring life. We should therefore keep in mind that the strategies we choose to reduce complexity, must always honour the eccentric, playful, 'under construction' nature of human life.

24 Mayer-Schönberger and Cukier 2013.
25 Dewandere 2015: 198.
26 Kahneman 2011.
27 Keymolen 2016.

28 Boyd and Crawford 2012: 666-668.
29 Dewandere 2015: 206.
30 Nissenbaum 2004: 173-174.

References

Aydin, C. and Verbeek, P. 2015. Transcendence in technology. *Techné: Research in Philosophy and Technology.* 19:3, 219-313.

Bednarz, J. 1984. Complexity and intersubjectivity: Towards the theory of Niklas Luhmann. *Human Studies.* 7:1, 55-69.

Boyd, D. and Crawford, K. 2012. Critical questions for big data: Provocations for a cultural, technological, and scholarly phenomenon. *Information, Communication & Society.* 15:5, 662-679.

de Mul, J. 2003. Digitally mediated (dis)embodiement. Plessner's concept of excentric positionality explained for cyborgs. *Information, Communication & Society.* 6:2, 247-266.

Dewandere, N. 2015. Rethinking the human condition in a hyperconnected era: Why freedom is not about sovereignty but about beginnings. *The Onlife Manifesto. Being human in a hyperconnected era,* edited by L. Floridi. 195-215. Cham: Springer.

Hildebrandt, M. 2015. The public(s) onlife. A call for legal protection by design. *The onlife manifesto. Being human in a hyperconnected era,* edited by L. Floridi. Charm: Springer. 181-194.

Ihde, D. 1990. *Technology and the lifeworld: From garden to earth.* Bloomington: Indiana University Press.

Kerr, I. and Earle, J. 2013. Prediction, preemption, presumption: How Big Data threatens big picture privacy. *Stanford Law Review Online.* 66, 65.

Keymolen, E. 2016. *Trust on the line. A philosophical exploration of trust in the networked era.* Forthcoming fall 2016: Wolf Legal Publisher.

Kitchin, R. 2014a. Big Data, new epistemologies and paradigm shifts. *Big Data & Society.* 1:1.

Kitchin, R. 2014b. Thinking critically about and researching algorithms. *The Programmable City Working Paper* 5, 1-19.

Luhmann, N. 1979. *Trust and power. Two works by Niklas Luhmann,* translated by H. Davis. New York: John Wiley & sons Ltd..

Mayer-Schönberger, V. and Cukier, K. 2013. *Big data: A revolution that will transform how we live, work, and think.* Boston: Houghton Mifflin Harcourt.

Nissenbaum, H. 2004. Will security enhance trust online, or supplant it? *Trust and distrust in organizations: Dilemmas and approaches,* edited by R. M. Kramer and K. S. Cook. 7. New York: Russell Sage Foundation.

Plessner, H. 1975. *Die Stufen des Organischen und der Mensch; Einleitung in die philosophische Anthropologie.* Berlin: De Gruyter.

Richards, Neil M. and King, Jonathan H. 2013. Three paradoxes of big data. *Stanford Law Review Online.* 66, 41.

Verbeek, P. 2010. *What things do: Philosophical reflections on technology, agency, and design.* Pennsylvania: Penn State Press.

Verbeek, P. 2011. *Moralizing technology: Understanding and designing the morality of things.* Chicago; London: The University of Chicago Press.

024 · Solace, 2011.

025 · Liquid Solace, 2015.

026 · Aurora, 2015.

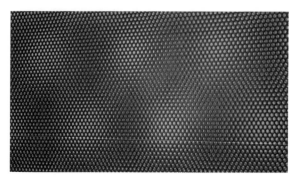

027 · Moiré Studies, 2013-2016, The moiré effect, which is strongly reminiscent of digitally created effects in computer graphics.

072

Aurora 2015.

029 · Artificial Ignorance software.

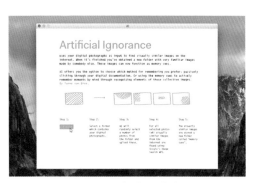

028 · Interface of computer application Artificial Ignorance.

042 · Tank_Rzl-Dzl-AI, The Innovation Engine.

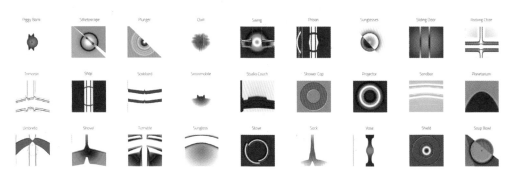

The Innovation Engine.

053 · Figure 1: The Boston Bombers.

079 · Lukang Longshan Temple. Lukang, Changhua, People's Republic of China. Courtesy of CyArk.org.

081 · The Black Obelisque, Assyrian Collection CyArk and the British Museum. Courtesy of CyArk.org.

080 · Isaac blessing Jacob; Catharijne Convent. Courtesy Wim van Eck.

UP

Facebook   Twitter   ...   Next Post

Trying to play Pokemon Go after decades of not leaving your house

101 • 9GAG post (12-08-2016).

Ongeveer 4.020.000 resultaten (0.39 seconden)

Could you survive a week with no internet? - Telegraph
www.telegraph.co.uk › Men › Thinking Man ▾ Vertaal deze pagina
8 jan. 2014 - It's nuts thinking that going 120 hours without the internet is an achievement, isn't it? Five
working days? Big deal. But let me explain. On a good ...

Killing Home Internet Is the Most Productive Thing I've Ever Done | T...
www.theminimalists.com/internet/ ▾ Vertaal deze pagina
I was not content with the time I was wasting—I felt I could do more purposeful things with my time than
spend it on the Internet. This doesn't mean I think the ...

30 Days without Internet - Planet of Success
www.planetofsuccess.com › Home › Conscious Living ▾ Vertaal deze pagina
18 feb. 2015 - Living a month without the internet would not have been difficult for a person that was living
200 years earlier. But for many of us who have ...

What would Life be Like Without the Internet? - Jamie King Media
jamieking.co.uk/lib/life-without-the-internet.html ▾ Vertaal deze pagina
How would having no Internet affect the general personal user? Well apart from every teenager screaming
because they can't log on to Facebook, it all comes ...

How To Survive With No Internet | SMOSH
www.smosh.com/smosh-pit/articles/how-survive-no-internet ▾ Vertaal deze pagina
27 dec. 2011 - If you don't have emotions at your disposal, there's no way anyone will understand the ...
Any other tips for surviving without the Internet?

I'm still here: back online after a year without the internet | The Verge
www.theverge.com/2013/.../im-still-here-back-online-after-a-year-without-the-interne... ▾
1 mei 2013 - One year ago I left the internet. ... With no clear idea how I did it, I wrote half my novel, and
turned in an essay nearly every week to The Verge.

How to Survive Without the Internet on Vacation: 10 Steps
www.wikihow.com › ... › Computers and Electronics › Internet ▾ Vertaal deze pagina
How to Survive Without the Internet on Vacation. Believe it or not, the internet has not always existed!
For those of you too young to know a time without internet ...

Can You Live Without An Internet Connection At Home? | Apartment ...
www.apartmenttherapy.com/can-you-live-without-an-intern-1311... ▾ Vertaal deze pagina
1 nov. 2010 - The best way to go online if you haven't got an Internet connection at home is ... As you
might have heard, Starbucks is no offering WiFi free for ...

Surviving without Internet: B.C. town offline since November living like...
news.nationalpost.com/.../surviving-without-internet-b-c-town-off... ▾ Vertaal deze pagina
6 jan. 2016 - With no Internet connection for almost six weeks, and the nearest bank four hours away,
Patricia Lynn, a district councillor in Stewart, B.C., has ...

102 • Google search 11-08-2016.

075

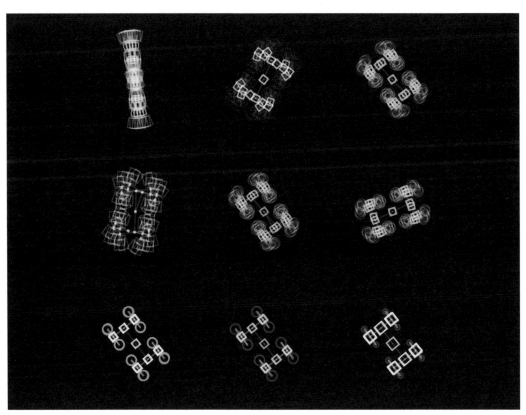

112 • Breeder 01.

112-113 · Breeder is a software application that provides its users the ability to playfully explore the principle of artificial evolution. The software is based on the 'Biomorph' program as proposed by the British evolutionary biologist Richard Dawkins in his book *The Blind Watchmaker*. Variables like the colors, patterns, and movement of abstract visual elements are encoded into an artificial DNA. The user can crossbreed and mutate the genetic codes of these elements, thus creating new generations. This leads to an endless stream of rhythmically pulsating images that highlight the poetic beauty of the evolutionary process.

Command Line

Graphical User Interface

Conversational Interface

132 · Amazon Echo. Image attribution: By Frmorrison at English Wikipedia, CC BY-SA 3.0, *commons.wikimedia. org/w/index.php?cu-rid=47040540.*

133 · Evolution from the command line to the voice interface.

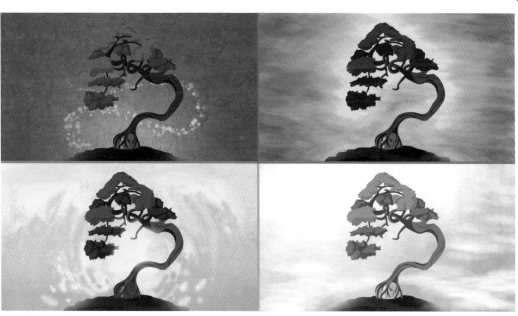

140 · Figure 2: Design of a life tree for an wisdom coach that supports end of life dialogues (design: GainPlay Studio, Utrecht).

DIGITAL CONSERVATION OF CULTURAL HERITAGE

Yolande Kolstee

Digital technology gives us a totally new perspective on the broadcasting of information and Internet has changed the way we communicate and share information in a way we could not have foreseen. Countless websites contain educational, artistic, scientific, and lifestyle information, next to websites offering entertainment in the broadest sense of the word. We can travel through time and space, see and hear things that we would never before have seen or heard. There are digital copies of fragile old books and manuscripts that we can look at, we can search for annotations and read what other people know or think about the pages we visit. We can educate others and ourselves by sharing knowledge via digitalised information, being text, images, film, or sound and music. This article aims to discuss various reasons for 'digitisation' and explain a few of the methods used. I shall illustrate this by showing a few examples from Europe and the United States. The underlying concepts of 'digital conservation' as well as 'cultural heritage' are complex. I will elaborate, very briefly, on these concepts to touch upon their complexity.

Digitisation and Digitalisation

Among the general public there is a reasonable understanding of the term 'digital': something that exists in a digital format - in the end 0s and 1s.[1] Concerning digitalisation, there is a difference between 'digitisation' and 'digitalisation'. The term 'digitise' refers to the material process of converting analogue information to a digital format; 'digitalise' refers to the adoption or increase of use of digital communication techniques, which we see in many, or almost all, domains of social life that are restructured or only exist around digital communication and media infrastructures.[2] In this article I will make use of both terms, digitisation - converting analogue to digital - and digitalisation – increasing use of digital media.

078   The conversion of an analogue artefact to a digital format has an immediate effect on its availability: once digitised, a 3D or 2D image can be distributed, customised, shared, changed, sold, lend out, exposed, appear on websites, in catalogues etc. I emphasise this aspect because digitising is not value-free. Bolter and Grusin discuss the consequences of remediation, and state that 'hypermedia applications are always explicit acts of remediation: they import earlier media into a digital space in order to critique and refashion them.'[3] In short, one question that arises is, 'what will be changed of the original artefact once digitised?' As stated by Cameron and Kenderdine in *Theorizing Digital Cultural Heritage, a critical Discourse*:

> Museums and heritage organizations have institutionalized authority to act as custodians of the past in Western societies. As such, they hold a significant part of the 'intellectual capital' of our information society. The use of emerging digital technologies to activate, engage, and transform this 'capital' is paralleled by shifts in the organizational and practice culture of the institutions entrusted with its care. In a symbiotic relationship, cultural heritage 'ecologies' also appropriate, adapt, incorporate, and transform the digital technologies they adopt.[4]

Cultural Heritage

When pondering the concept of 'cultural heritage' we enter a domain with many definitions, narrow ones and very broad ones, that are embedded in various disciplines. The meaning of the term 'cultural heritage' has changed considerably in recent decades and the concept is still in a developing stage. Cultural heritage does not just:

> [...] encompass monuments and collections of objects. It also includes traditions, or living expressions, inherited from ancestors long gone and passed on to their descendants, such as oral traditions, performing arts, social practices, rituals, festive events, knowledge and practices

1,2 Brennen and Kreiss 2014. 'Scholars across disciplines use the term digitization to refer to the technical process of converting streams of analogue information into digital bits of 1s and 0s with discrete and discontinuous values. Feldman (1997) argues, unlike analogue data with 'continuously varying values, digital information is based on just two distinct states. In the digital world, things are there or not there, 'on' or 'off'. There are no in-betweens.' That digital bits have only two possible values leaves many to argue that, in the words of Pepperell: 2003 'digital information is discrete and 'clean', whilst analogue information is continuous and 'noisy'.'
3 Bolter and Grusin 1996: 311-358.
4 Cole 2008: 425-426.

concerning nature and the universe, or the knowledge and skills to produce traditional crafts.[5]

The term that covers this type of heritage is 'intangible cultural heritage', yet, in this article I will concentrate on tangible cultural heritage.

Cultural heritage includes material and intangible aspects of human life. UNESCO, acronym of United Nations Educational, Scientific and Cultural Organization, is an important player in the field of cultural heritage. It has, among many others, several committees for aspects of *nature* labelled as 'cultural heritage'- think of cultural landscapes or gardens that were built upon e.g. the philosophy of the Freemasonry, or typical gardens as part of the life of Roman Catholic monks or sisters, or pilgrimages and other cultural routes. Knowledge of the religious, sacred, or spiritual meaning of cultural sites in nature, as well as the artistic meaning is essential to estimate the value of those forms of cultural heritage.

We must keep in mind that as soon as we want to add 'meaning' to a cultural heritage, we enter the world of narratives, and with the narrative inextricably linked to the physical or digitised object, we have then entered the world of non-tangible heritage. With these concepts as the basis for this small exploration, I would like to draw on a few examples to give a view of the field of 'digital conservation of cultural heritage'. Although first, why would we digitally preserve cultural heritage anyway?

Reasons for Converting Material Objects or Sites
There are many reasons (that sometimes overlap) for converting non-digital material or music to a digital format.

Protection, Preservation and Safeguarding
The UNESCO declaration states that protection and safeguarding of world heritage from natural disasters and disasters caused by human conflict, is needed because 'World Heritage sites belong to all the peoples of the world, irrespective of the territory on which they are located.'[6]

Research
Digitising artefacts like paintings, pottery, statues and even landscapes, makes it possible to execute detailed research by many scholars at the same time. Researchers can cooperate, even if research-centres are located geographically far from each other.

Increase Access and New Forms of Access for New Audiences
Digital access creates opportunities for life-long learning, for educating schoolchildren and for anyone who does not have the means to visit distant temples, libraries, or museums. For example, the digitising of ancient texts from the Middle East, formerly only accessible for visitors in Situ, gives people anywhere at any time the possibility to study, or just marvel at, the texts. Annotating the sight or texts with AR or VR can provide context needed to understand their significance.

Reduction of Handling
Some fragile artefacts need to remain in conditioned circumstances to preserve them, and only a limited amount of people is allowed to touch and study them. When digitised, those artefacts can be closely observed without the danger of damaging them. This is for 3D objects the same as for texts and manuscripts, photos and pictures as well as for moving images as film and any other visual or auditory material.   079 · Lukang Longshan Temple. Lukang, Changhua, People's Republic of China. Courtesy of CyArk.org.

Create Collections of a Specific Type of Artefact
With the digitised objects it is possible to construct a (private) digital collection or catalogue of objects or texts. Before digitisation, if one had an interest in, for example, European pottery, one had to visit museums in different countries or even continents (when artefacts were, temporarily, outside Europe).

5 Centre 1992.
6 Centre 1992. UNESCO seeks to encourage the identification protection and preservation of cultural and natural heritage around the world considered to be of outstanding value to humanity. This is embodied in an international treaty called The Convention concerning the Protection of the World Cultural and Natural Heritage adopted by UNESCO in 1972.

Methods of Digitising

When scanning an object, you (re)produce an electronic, digital form. There are many specialist multi-spectral image-processing techniques such as X-Ray, X-RF (fluorescence imaging), infrared and ultra-violet spectroscopy. While using these techniques, one irradiates the object with special wave-lengths of light and the reflected radiation will be recorded, or transmitted, by a special camera that is sensitive to this area in the light-wave spectrum.

When making a 3D scan of an object, a laser scanner bounces light off an object and collects the resulting topological cloud of points. To reproduce every nook and cranny, the scanner snaps overlapping images from all possible angles. A computer then sews together one large surface image and draws lines from one point to another to create a wire-frame model. High-resolution digital cameras add color and texture. When fully assembled, the models can be viewed, printed or manipulated. These scans do more than pickle a memory in a database. […] And when a site is destroyed, the scans can even be used to reconstruct what was there. That has already happened for one World Heritage Site, the Kasubi Tombs in Uganda. Built of wood in 1882, they were destroyed by fire in 2010 and rebuilt in 2014, based in large part on 3-D models made in 2009.[7]

080 · Making captures in The Mauritshuis- Rembrandtproject. Courtesy prof. P.P. Jonker.

Examples of Digital Conservation of Cultural Heritage

080

A very spectacular form is used by the CyArk foundation, which specialises in making digital 3D models of huge sites and sculptures.[8]

3D scanning of paintings **074**

080 · Isaac blessing Jacob; Catharijne Convent. Courtesy Wim van Eck.

Due to the innovative work done by professor Pieter Jonker from the vision-based robotics group at the TU-Delft, and notably his PhD student Tim Zaman, high-resolution 3D scanning techniques have improved. His work on simultaneously topographic and colour scanning for Art and Archaeology was presented at the Technart16 conference at the Rijksmuseum in Amsterdam. In close collaboration with professor Joris Dik, a new scanning technique was developed that scans the surface of paintings in three dimensions, (x, y, z) with a resolution of ten microns. Aside from digitisation for monitoring paintings in 3D over time, digital image processing for art historians and material scientists, 3D printing of paintings is a goal of this scanning technique project. The 3D printing of the paintings was achieved by professor Jo Geraedts from Océ / TU-Deft and his team. They attracted national and international attention with their 3D print of The Jewish Bride and Self-portrait by Rembrandt and flowers by Van Gogh. These 3D-printed paintings were on show during the conference.[9]

7 EDITORS 2016.
8 Barton 2010. Surveying techniques range from simple manual methods to remote sensing instruments capable of millimetre precision. The choice of survey device depends on the type of information needed and the corresponding degree of accuracy that is required. Combined with the use of a clinometer (vertical angle measurement) and compass (horizontal angle), or theodolite (vertical and horizontal) readings, three dimensional coordinate information can be deduced with a reasonable level of accuracy. Total stations provide angle and distance measurement devices within one instrument, thus enabling the calculation of 3D coordinates without the need for other measurements. Total stations enable distance calculation methods using remote sensing devices typically capable of longer-range measurement than manual devices.
9 Kolstee 2013. This paragraph already appeared in 'AR and cultural heritage: A healthy and mature relationship'.

Interactive installation in Museum Catharijne Convent, Utrecht, The Netherlands, where the public, via a tablet, could find information based on multi-spectral images of the painting 'Isaac blessing Jacob' by Govert Flinck (1615-1660).[10]

A special technique to show results of research is augmented reality (AR). AR is a non-invasive technique, meaning that an object or site will not be touched or corrupted, whilst enabling visitors to engage with artefacts, or using devices that are in no way physically familiar or connected to them. Besides the non-invasive character of AR, there is also another important aspect to consider: The sanctity or holiness of artefacts (as perceived by some) makes it impossible, or at least extremely disrespectful, to touch and manipulate the objects. Yet, in AR, this is possible in virtual space.

In the case of non-holy objects, the delicacy of artefacts might call for a strict do-not-touch policy. With AR, however, you can see in real-time, for instance, the back of a priceless painting projected upon its surface or a painting *underneath* the one that we can see on the canvas. The Van Gogh Museum in Amsterdam, The Netherlands, dedicated an exhibition to the 'Reuse of canvases of Van Gogh' in 2011 and used AR to show the research on these paintings. More cases will follow, because research in museums has revealed scenes beneath the painted scene we see when looking at the work, e.g. work of Picasso, Hendrik van Anthonissen, Leonardo da Vinci, Giampieta Campana.[11]

Rohzen Mohammed-Amin states another reason to use AR: 'with AR we can show objects or sites that no longer exist. Sometimes this is due to natural disasters such as earthquakes in 2010 and 2011, as happened in Christchurch, New Zealand, in which case there are countless photographs and other 'souvenirs' that can help to revive the demolished city in AR.'[12] By working closely with the CityViewAR12 project team at HITLabNZ13, Mohammed-Amin observed how AR helped to recreate views for the people of Christchurch who had to say goodbye to Christchurch's landmark buildings.[13]

However, objections have been raised by theorists –among many other aspects- to prevent a possible 'Disneyfication' of cultural heritage, by making too much use of interactive installations in museums to raise the 'fun-factor'.[14]

Vision

Digital conservation of cultural heritage is important for safeguarding, research, education, preservation, and virtual exchange of historical artefacts. Various scanning methods make it possible to compare artworks, to learn about the materials used and the construction of the artefacts. Techniques like AR and VR make it possible to annotate in an interactive way. UNESCO's vision is that:

> [...] upholding strict guidelines is necessary to protect culture, nature, and cultural objects, buildings, and archaeological sites. In today's interconnected world, culture's power to transform societies is clear. Its diverse manifestations – from our cherished historic monuments and museums to traditional practices and contemporary art forms – enrich our everyday lives in countless ways. Heritage constitutes a source of identity and cohesion for communities disrupted by bewildering change and economic instability. Creativity contributes to building open, inclusive and pluralistic societies. Both heritage and creativity lay the foundations for vibrant, innovative and prosperous knowledge societies.[15]

By converting physical material to a digital format and by making the results available through the Internet, we make our cultural heritage available to anyone, anywhere, whether it is solely for taking pleasure in esthetical sense but also for educational and research purposes. In doing so we can contribute to a world in transition in which sharing, co-creating, and mutual responsibility will overcome exclusiveness and loss of identity caused by inaccessibility of artefacts of our own or our neighbours' roots. Theorists in conferences on this issued discuss the possible effects and potential downsides to digitisation.[16]

081 • The Black Obelisque, Assyrian Collection CyArk and the British Museum. Courtesy of CyArk.org.

10 Museum Catharijne Convent.
11 Campbell-Dollagham 2014.
12 Kolstee 2013. This paragraph already appeared in 'AR and cultural heritage: A healthy and mature relationship'.
13 At your fingertips: What Christchurch looked like - business - NZ herald news: 2012.
14 Cole 2008: 425-426.

15 UNESCO 2016.
16 Conferences such as Museums and the Web, International Cultural Heritage Informatics Meeting, International Symposium on Virtual Reality, Archaeology and Intelligent Cultural Heritage, Computing Archaeology, Virtual Systems and Multimedia, Electronic Visualization and the Arts.

References

Barton, J. 2010. 'CyArk'. 1 January. Accessed 20 September 2016 at:
  cyark.org/education/surveying.

Bolter, J.D. and Grusin, R.A. 1996. *Remediation.* Georgia Institute of Technology 4:3, 311–
  358.

Booker, J. 2012. At your fingertips: What Christchurch looked like. *NZ herald
  news.* Accessed 20 September 2016 at:
  www.nzherald.co.nz/business/news/article.cfm?c_id=3&objectid=10814442

Brennen, S. and Kreiss, D. 2014. *Digitalization and Digitization,* 8 September. Accessed 20
  September 2016 at: culturedigitally.org/2014/09/digitalization-and-digitization/.

Campbell-Dollaghan, K. 2014. 5 lost images found hidden beneath famous paintings.
  Accessed 20 September at: gizmodo.com/5-lost-images-found-hidden-beneath-famous-paint-
  ings-1592796080.

Centre, U.W.H. 1992. UNESCO world heritage centre - frequently asked questions (FAQ) -
  intangible heritage. Accessed 20 September 2016 at: whc.unesco.org/en/faq/40.

Centre, U.W.H. 1992. *World heritage.* Accessed 20 September 2016 at: whc.unesco.org/en/about/

Cole, F. in Cameron, F. and Kenderdine, F. (Ed.) 2007. *Theorizing digital cultural heritage: A
  critical discourse. Theorizing digital cultural heritage: A critical discourse.*
  Cambridge, MA: The MIT press 200. 425–426.

EDITORS, T. 2016. 3-D digital modeling can preserve endangered historic sites forever.
  Arts & Culture in Scientific American, 1 July 2016.

Feldman, T. 1997. *An introduction to digital media.* London: Routledge.

Kolstee, Y. 2013. AR and cultural heritage: A healthy and mature relationship. *AR[t]
  Augmented Reality, Art and Technology,* 4 November 2013. Available at:
  https://issuu.com/arlab/docs/art4_issuu.

Museum Catharijne Convent. Invitation to presentation of exhibition: A world first: revealing
  a painter's secrets Govert Flinck's Isaac blessing Jacob unravelled. May 2013.
  Accessed 20 September 2016 at:
  www.codart.nl/images/Govert_Flinck_Catharijneconvent.pdf.

UNESCO 2016. Protecting our heritage and fostering creativity. Accessed 20 September
  2016 at: en.unesco.org/themes/protecting-our-heritage-and-fostering-creativity.

PASTS, PRESENTS, AND PROPHECIES ON YOUR LIFE STORY AND THE (RE)COLLECTION AND FUTURE USE OF YOUR DATA

Marjolein Lanzing

It's giving over to the technology this power to tell you who you are, to guide how you live.

Natasha Singer[1]

The world is becoming increasingly enchanted. It is no longer science fiction to imagine a world in which my espresso-machine will communicate with my alarm clock to ensure I will get up at 7:00 AM by being lured into the kitchen by the stimulating smell of coffee, while the Nano-chips in my body -in consultation with my fridge and calendar of course- decide whether to make an appointment with my local gym or to order some of my favourite chocolate. It saves time now that companies can predict what kind of products I will (or should) need: a nice outfit for a romantic date, anti-histamines, the latest novel by Javier Marias, diapers, or camping gear. At last, I have some peace of millennial mind now that algorithms that provide tailor-made options based on my informational history have tackled my choice anxiety.

Big Data is the revolutionary promise that by aggregating and combining enormous sets of data we can detect patterns that will subsequently allow us to predict and anticipate on future events, behaviour, or needs.[2] The lifeblood of the enchanted world is personal information and its increased collection, analysis, and combination.

This essay questions the increasing gathering of personal information by a host of 'smart objects' and the subsequent creation of more precise and more complete digital records of persons from the perspective of narrative identity.[3] In particular, it will explore the future challenge of profiles and their predictive power for one's personal identity based on these records. While – autobiographical - memory is valuable for identity, forgetting is equally important. A digital record may haunt us in the future in a way that we cannot possibly imagine or foresee today. If we want to remain dynamic and evolving selves with an open, self-chosen future, we should care about what should be deleted, remembered, and accessed.

Delete me (Not)

More and more of our data is collected and stored. Anyone who uses the Internet has been contributing to an online archive. This archive is an extended, digital memory that may become more complete and even hyperaccurate when more data of all your activities will be logged and stored.[4] In the future, it may amount to constituting a virtual doppelganger, a data mirror 'self'. Because deleting, forgetting, and filtering are time consuming and costly practices, the default is 'remembering'. The question is whether this bias affects our identity in a significant way.

Let us first take a look at the value of memory for the individual. Memory has an important social and psychological value. The social value of memory is straightforward: memory enables us to hold each other accountable and it causes us to act responsibly in society. If we know that our actions and decisions are remembered and could be rubbed in our faces later in life, we will think twice before we act.

The psychological value of memory becomes apparent when we examine a pathological case in which memory is completely absent. In the case of amnesia, total or partial loss of memory, we usually recognise that this hurts someone's personality and sense of selfhood. People who suffer from amnesia, dementia, or Alzheimer's disease are not able to coherently construct their life-stories, even to the extent that they do not recognise their spouses or children anymore.

1 Singer 2015.
2 boyd & Crawford 2011: 2.
3 Smart objects are objects that are aware of their surroundings and can react to activities and events that occur in the physical world. Their abilities depend on the way they gather, process and analyse (personal) information. This also enables them to interact with,

for instance, humans or other smart objects.
4 For the first 'lifelogger', see: Bell, G. & Gemmell, J. 2010. 'Your Life Uploaded'. New York: Plume. Gordon Bell was the first person to actively pursue an extended memory by digitalising his entire life.

Remember the jarring scene from *The Notebook* in which Allie panics when her husband tries to console her because she does not recognise him. In the course of their life Allie has written the story of their love and it is only when her husband reads this story to her that she, however briefly, regains her memory and becomes 'Allie' once more.

Memory is valuable for individuals in society, but so is forgetfulness. Without forgetfulness, society would become a suffocating place and so, 'social forgetfulness' has even been institutionalised in domains such as credit history and juvenile criminal behaviour.[5] Forgiveness and forgetting are very closely related and social forgetting may allow people to try again without the constant burden of their past. To be freed from one's past mistakes and to be granted a clean slate may cause people to feel trusted and to reinstate trust in themselves. We trust that (some) people learn from their mistakes and should have the chance to start over with all their options open.[6] This was also reflected in the court ruling by the European Court of Justice in the case of Google Spain v AEPD and Mario Costeja González, which amounted to an renewed interest in and application of the 'Right to be Forgotten'.[7] From now on search engines must remove links to data if a person finds that the data the link refers to is 'inadequate, irrelevant or no longer relevant, or excessive in relation to [the] purposes [of the processing] and in the light of the time that has elapsed.'[8]

While remembering is important for understanding 'who we are', there is also psychological value in forgetting.[9] Total recall can be just as problematic for one's sense of self as memory loss can be. Hyperthymesia, the condition of suffering from an extremely detailed autobiographical memory, can cause one's memory to become tyrannical and prevent a person from constructing their life stories.[10] Think about people who went through traumatic experiences and need to be able to 'forgive and forget' others or themselves. Reconstructing our memories would help us by allowing us to reshape and reinterpret past experiences and to create a personal narrative that tells us, and others, who we are. This would enable us to move on and to make room for new memories. 'Blessed are the forgetful, for they get the better even of their blunders.'[11]

Autobiography

Having established the social and psychological value of remembering and forgetting, we can now explore how that relates to a 'healthy' or dynamic identity. The latter I define as an identity that is fluid and can evolve or change over time. A dynamic identity requires an autonomous agent and an open future. Along with Burkell, I will refer to such an identity as a narrative self: someone who can understand herself as existing across time and who can own her experiences from the past as '*hers*' – as opposed to the idea that 'the person who once made that decision is not the person I am today and therefore I cannot say that choosing to study philosophy in 2007 was '*my*' decision.'[12] That means that every experience will need to be interpreted and integrated into someone's life story. When I tell people how I ended up practising philosophy, it will be *my* story precisely because it is *me* who retrieves, interprets, arranges and reconstructs the memories in a way that is consistent, or in coherence with, my life projects.

In the amnesia-case we saw that being able to narrate your own life story is important for one's sense of self and, therefore, memories are also important for one's sense of self.[13] Although memory is very important to a narrative identity, we also established that a totally objective, accurate, and complete memory would not serve it well. The human mind is patchy, fallible, and subjective but that may be for the better since, we forget, select, alter, polish, emphasise, and confabulate events in retrospect.[14] It is not the exact documentation of our lives that tells us who we are, it is the way *we* reconstruct and tell the story. This allows us to learn from our mistakes, to grow, to develop, to forgive, and to redeem.

Forgiving and forgetting are important for a subjective narrative. Without it, it would be

5 Blanchette and Johnsson 2002.
6 Of course, we are particularly willing to forgive and forget childhood.
7 Harvard Law Review December 2014 Case C-131/12; European Court of Justice 13 May 2014, C-131/12, paragraph 93.
8 Harvard Law Review December 2014 Case C-131/12; European Court of Justice 13 May 2014, C-131/12, paragraph 93.
9 For extensive research on identity and the psychological necessity of forgetting, see Burkell 2016.
10 Burkell 2016: 18.

11 Nietzsche 1999 [1886]: 129 Ch. 7 Aph. 217.
12 Schechtman 1996; Burkell 2016: 17-18.
13 There are many different types of memories such as skills (riding a bicycle) and 'facts' (my mother's birthday). However, I am interested in narrative autobiographical memories or storylines that stretch across a longer period of time. See also Christman 2011: 88.
14 Of course, there is also a limit to the subjectivity of memories. Memories are socially constituted: we need the corroboration of our peers. We cannot simply 'imagine' past experiences (then they would also no longer count as memories of course).

very difficult to learn from and develop our selves beyond our past actions. Instead, we would 'freeze' our identities by the eternal memory of our past actions and foreclose the possibility of change. Regrets, constant analysis of the past, an inability to let go of former experiences or a constant longing for the past can sabotage one's ability to engage in new experiences. While responsibility is obviously necessary in order to be an agent, we lose a certain degree of spontaneity, experimentation, freedom, and creativity when we know that everything is remembered and that we can be confronted with these memories. This can only be a loss for a democratic society that thrives on free agents who uphold different and conflicting conceptions of the good life.

Profiles, Targets and Predictors

How does a future hyperaccurate memory of you affect your identity? In order to fully grasp the implications and the future challenges, we first need to map its role within the promise of Big Data. Every time you use your smartphone, surf the web, click on an advertisement, post a comment, picture or video, track your sleep, moods or runs, the digital record of 'you' will become more detailed. Your behavioural patterns in combination with the aggregated footprints you and many others leave behind in the online forest will form a certain trail or 'profile'.

'Profiling' is the – social – categorisation of users based on patterns detected within these mined data aggregates.[15] These patterns are then representations and means of identification of persons. Profiling is a very lucrative business because it enables behavioural targeting and personalised advertising. The profiles can be used to predict whether someone, for instance, will be a costly health insurance client or whether someone is likely to repay a student loan in the future.[16] Also, based on profiles, advertisements can be efficiently directed at people who will most likely buy certain products. This is called targeting. For instance, a very common experience among females between 25 and 35 is that they are targeted with advertisements related to pregnancy or maternity.

The promise of Big Data lies in the realm of predictive analysis:

> Predictions based on correlations lie at the heart of big data. Correlation analyses are now used so frequently that we sometimes fail to appreciate the inroads they have made and the uses will only increase.[17]

The quantity and granularity of information is growing every day: we not only have access to more information, but we can also retrieve more different kinds of information such as geo-location, content, duration, ip-addresses, biometrics etc. In addition, we have unprecedented access to information because sharing and spreading information has never been easier – this in sharp contrast with removing or deleting information, which is costly, time consuming and virtually impossible. Also, our abilities in combining and correlating data have advanced. Together, these technologies have predictive power. By combining past and current data, it will be possible to create profiles based on newly discovered patterns of behaviour and to generate new information that was not available in the original data sets.[18] It will be possible to generate information about individuals or groups that they did not know about before themselves. It might then be possible to predict future actions based on these behavioural patterns and to anticipate future choices.

The power of prediction that is present in large aggregates of personal data first became apparent with the Target-case. The retail company Target learned – through the data collected via the discount card – that a girl had switched her regular shampoo for a brand of which the scent was popular among pregnant women. Consequently, Target started sending discount coupons for diapers to the family. When the – rather upset - father asked Target to explain the coupons, and they explained the altered shopping pattern, he learned of his teenage daughter's pregnancy. Although the predictive powers of Target have been disputed, it remains an impressive case that makes you wonder about the future possibilities of Big Data and its predictive powers.[19] Amazon already patented 'anticipatory shipping' in 2014. Based on one's data, the retail company will ship the product even before the customer has clicked on the 'buy' button.

15 Hildebrandt and Gutwirth 2008: 2.
16 For extensive and detailed research on profiling, see Hildebrandt and Gutwirth 2008.
17 Mayer Schonberger & Cukier 2013: 55-56.
18 Blanchette and Johnson 2002: 39.
19 Siegel 2014; Duhigg 2012.

If it will be possible to anticipate what people will be likely to buy in the future, it might not be a sci-fi scenario to explore the other realms in which predictions can be made. Predictive analysis will provide us with new understanding about who we are but also about who we will be. The question is whether this benefits a dynamic identity.

Freezing and Recycling

There are several challenges for a dynamic identity that seem to arise from digital archives, profiles, and their predictive powers. First of all, there is something problematic about profiles as social categorisations and the self-fulfilling prophecies they can become. By repeatedly offering – future– choices and options based on past behaviour, profiles might be prone to a looping effect. Recycling may actually fix a person's identity and reinforce the profile or category. This may also cause a kind of tunnel vision or filter bubble in which the options are tailor-made, but close alternatives would not fit the profile. Even though the choices that are offered to us based on our past might have been the choices we *would* have made -even if there would have been a thousand other options- is beside the point. The point is that we want to make these choices ourselves, and go through the process of choosing from various options. This is what makes our decisions authentic and what makes our life story a personal one. Without that process, we are no longer the voices or narrators of our own stories. This may not contribute to a dynamic self with a personal narrative about her self-development. Worst-case scenario: people will start displaying behaviour that corresponds with their profiles, following their profile-based predictions, thus 'becoming who they are' – steered by and according to the digital archive.[20]

Relatedly, 'freezing' a person by recycling past behaviour deprives persons of an open future. Recycling does not take into account a person's change over time and it fixes the future based on automatically selected snippets of information from the archive.[21]

This brings me to third point, which is the fact that while a dynamic, narrative identity requires an autonomous agent who tells her personal life story, we will lose control over our personal narratives when our stories are 'told' by profiles and their automatically generated predictions about our future behaviour. The possibilities for our own personal interpretation and subjective reconstruction of events, and so the possibilities for constructing a personal narrative, will be undermined and might even become impossible or unwanted. Especially due to the aura of 'factuality', 'objectivity' and 'neutrality', which seems to surround the numbers in data science and analysis. As Hildebrandt argued:

> [...] the characterised identity held by a computer will override any other potential information, including apparent visual ones. We are witnessing that our evolved dominant (visual) sense is being superseded by computerbased profiling information.[22]

Finally, a permanent digital archive will have a self-disciplining or Panoptic effect. Instead of experimenting and being creative in our actions and decisions, the fact that all will be saved and *could* come to haunt you at later stage, during a job interview for instance, may actually petrify you. If digital forgetfulness or forgiveness does not exist, we will become afraid to make mistakes. After all, we cannot reconfigure these mistakes as part of a learning process and as part of our narrative identity, because the digital record mercilessly presents the 'facts' and presents the verdict about 'future you' based on who you once were or what you once did.

Moving On

It is the first-person point of view and interpretation of one's experiences that contributes to one's self-understanding. By looking through the ever-changing framework of my personal values, beliefs, and desires that I accrued over time, I can distance myself from past experiences and embrace others. This is the outlook I use to construct my personal life story. Currently, we are contributing to giant personal databases and we do not know what this information can and will be used for in the future. However, if predictive analysis takes a flight, we should be prepared to

20 Hacking 2006.
21 Allen 2008: 64.
22 Hildebrandt and Gutwirth 2008: 340.

re-encounter the information we contribute today in a future form we cannot possibly foresee at this point. If we care about the freedom and autonomy of our future selves, then we should at least design digital expiration dates for categories of information that we think have a right to be forgotten, make the rationales behind profiles transparent and control the information of our digital archives. After all, in order to have an open future we should be free to tell our own stories and move on.

References

Addis, D.R. and Tippett, L.J. 2004. Memory of myself: Autobiographical memory
    and identity in Alzheimer's disease. *Memory.* 12:1, 56-74.

Allen, A. L. 2008. Dredging up the past: Lifelogging, memory, and surveillance.
    The University of Chicago Law Review. 75:1, 47-74.

Blanchette, J-F and Johnson, D.B. 2002. Data Retention and the Panoptic Society:
    The Social Benefits of Forgetfulness. *The Information Society.* 18, 33-45.

boyd, d. & Crawford, K. 2012. Critical questions for big data: Provocations for a
    cultural, technological, and scholarly phenomenon. *Information, Communication & Society.* 15:5,
    662-679.

Burkell, J. 2016. Remembering me: big data, individual identity and the
    psychological necessity of forgetting. *Ethics and Information Technology.*
    18:1, 17-23.

Christman, J. 2011. Memory, Agency and the Self. In: *The Politics of Persons:
    Individual Autonomy and Socio-Historical Selves.* Cambridge: Cambridge University Press.

Dodge M., and Kitchin, R. 2007. Outlines of a world coming into existence:
    pervasive computing and the ethics of forgetting. *Environment and Planning B: Planning and De-
    sign.* 34:3, 431 – 445.

Duhigg, C. 2012. How Companies Learn Your Secrets, *The New York Times*
    Magazine. 16 February 2012 at: *www.nytimes.com/2012/02/19/magazine/shopping-habits.html.*

Hacking, I. 2006. Making Up People: clinical classifications. *London Review of
    Books.* 28:16, 17 August 2006, 23-26. Accessed at: *www.lrb.co.uk/v28/n16/contents.*

Harvard Law Review December 2014. Cyberlaw/Internet Case C-131/12. Accessed at:
    harvardlawreview.org/2014/12/google-spain-sl-v-agencia-espanola-de-proteccion-de-datos.

Mayer-Schönberger, V. and Cukier, K. 2013. *Big data: A revolution that will
    transform how we live, work, and think.* Boston: Houghton Mifflin Harcourt.

Nietzsche, F. 1999 [1886]. Voorbij Goed en Kwaad. Amsterdam: Uitgeverij de Arbeiderspers.

Schechtman, M. 1996. The Constitution of Selves. Ithaca NY: Cornell University Press.

Siegel, E. 2014. Did Target Really Predict a Teen's Pregnancy? *KD Nuggets News.* Accessed at:
    www.kdnuggets.com/2014/05/target-predict-teen-pregnancy-inside-story.html.

CAN CONNECTED MACHINES LEARN TO BEHAVE ETHICALLY?

Ben van Lier

Over the past few years, the rapid development of artificial intelligence, the huge volume of data available in the cloud, and machines' and software's increasing capacity for learning have prompted an ever more widespread debate on the social consequences of these developments. Autonomous cars or the application of autonomous weapon systems that operate based on self-learning software without human intervention, are deemed capable of making life-and-death decisions, are leading to questions on a wider scale about whether we as human beings will be able to control this kind of intelligence, autonomy, and interconnected machines. According to Basl, these developments mean 'ethical cognition itself must be taken as a subject matter of engineering.'[1] At present, contemporary forms of artificial intelligence, or in the words of Barrat, 'the ability to solve problems, learn, and take effective, humanlike actions, in a variety of environments,'[2] do not yet possess an autonomous moral status or ability to reason. At the same time, it is still unclear which basic features could be exploited in shaping an autonomous moral status for these intelligent systems. For learning and intelligent machines to develop ethical cognition, feedback loops would have to be inserted between the autonomous and intelligent systems. Feedback may help these machines learn behaviour that fits within an ethical framework that is yet to be developed.

Industrial Revolution

In the middle of the 20th century, the first major steps were made in the area of system theory, cybernetics, and computer science. This has led to the creation of things such as telecommunications, personal computers, and software. Together, these technological advances provide huge volumes of data and information in our modern world. By the end of the 20th century, these developments had already led Weiser to predict that, following the interconnection of people and computers through the internet, this interconnectedness will form a new basis for new technological applications.[3] New and internet-based applications would, according to Weiser, crop up within a decade and herald a new wave of technological possibilities that will lead to 'ubiquitous computing', which he defined as: 'the connection of things in the world with computation. This will take place at many scales including the microscopic.'[4]

This development of interconnecting objects in networks such as the internet is, however, moving ahead so rapidly that it led Greenfield to assert that, 'when everyday things are endowed with the ability to sense their environment, store metadata reflecting their own provenance, location, status and use history, and share that information with other such objects, this cannot help but redefine our relationships with such things.'[5] Greenfield's observation of the new relationship between humans and objects, and how this relationship changes our everyday reality, is becoming an increasingly topical one. Ten years after Greenfield's assertion, developments such as the smartphone, laptop computer, tablet computer, or sensors (you can even get trainers with sensors these days) are slowly but surely becoming normal elements of our everyday living and working environment. Earlier this year, the chairman of the World Economic Forum therefore rightfully said that 'It began at the turn of the century and builds on the digital revolution. It is characterized by a much more ubiquitous and mobile Internet, by smaller and more powerful sensors that have become cheaper, and by artificial intelligence and machine learning.'[6] In his view, new possibilities are continuing to emerge in this day and age that will allow us, and even make it second nature for us, to interact with technological applications such as cars, television sets, smartphones, or tablet computers on an increasing scale. In his opinion, these developments will spawn a kind of ubiquitous computing where 'robotic personal assistants are constantly available to take notes and respond to user queries'. For Schwab, however, it is not only about increasing interconnectedness between objects and objects' increasing

1 Basl 2014.
2 Barrat 2013: 25.
3 Weiser 1996.

4 Weiser 1996: 4.
5 Greenfield 2006: 23.
6 Schwab 2016: 7.

intelligence. He claims that it is important now to assume a broader view and contextualise the aforementioned development alongside contemporary technological developments such as DNA sequencing, nanotechnology, and quantum computing. In Schwab's view, these new combinations have the power to further strengthen the on-going process of networked people and objects blending together. This will produce revolutionary changes over the coming years, 'It is the fusion of these technologies and their interaction across the physical, digital and biological domains that make the fourth industrial revolution fundamentally different from previous revolutions.'[7] Brynjolfsson and McAfee have also seen growth in the application of new combined possibilities of hardware and software in global networks, this signals that scale is losing importance for the resulting new applications. They claim that we are gradually coming to an 'inflection point – a point where the curve starts to bend a lot – because of computers. We are entering a second machine age'[8] In this second machine age, man and machine will, in their view, become better able to use the huge variety of data and information they produce and consume together. The real benefits of this age will, according to them, only be visible when man and machine are sufficiently able to autonomously seize the opportunities offered by the data and information in terms of product and service development. In the words of Brynjolfsson and McAfee, 'countless instances of machine intelligence and billions of interconnected brains working together to better understand and improve our world. It will make a mockery out of all that came before.'[9] Given the fact that humans and machines will increasingly work together in networks, Brynjolfsson and McAfee feel we have to ask ourselves, 'what is it what we really want and what we value, both as individuals and as a society.'[10]

Increasing interconnectedness of technological applications and their ever greater intelligence are reflected in developments such as self-driving cars, drones, smart TVs, and thermostats that are installed in millions of homes and share energy consumption data with their environment. General Electric's Evans and Annunziata claim that this interconnectedness of machines in networks and their growing intelligence is bound to also lead to them rapidly developing an autonomous ability to make decisions. They claim that 'once an increasing number of machines are connected within a system, the result is a continuously 'Once expanding, self-learning system that grows smarter over time.'[11] When machines have autonomous access to data and information and are able to use it to learn from their behaviour, they will also gradually become able to acquire functionality that will enable them to take on tasks that are currently handled by human operators. According to Evans and Annunziata, this transfer of tasks from humans to machines 'is essential to grapple with the increasing complexity of interconnected machines, facilities, fleets and networks.'[12] Their growing capacity for learning enables machines to make decisions together that intervene in our lives and alter our reality without us noticing. According to Arthur, the increasing interconnectedness between man and technology is one of the drivers behind a development that is shaping our world within a 'network of functionalities - a metabolism of things-executing things – that can sense its environment and reconfigure its actions to execute appropriately.'[13] He says that 'We are beginning to appreciate that technology is as much metabolism as mechanism': comparing it to biological systems.[14]

Artificial Intelligence

Interconnected machines are able to become smarter and acquire greater capacity for learning thanks to developments in the area of data analysis, deep learning, and neural networks. The development of Artificial Intelligence has the same scientific roots as that of the computer. In 1950, Claude Shannon wrote that 'modern general-purpose computers can be used to play a tolerably good game of chess by the use of suitable computing routine or 'program'.'[15] At roughly the same time Turing asked himself the question, 'can machines think?'[16] Both Shannon and Turing were convinced at the time that widespread research would be required in years to come to develop algorithms and software programs to make it possible to have a computer answer these ambitious questions.

7 Schwab 2016: 8.
8 Brynjolfsson and McAfee 2014: 9.
9 Brynjolfsson and McAfee 2014: 96.
10 Brynjolfsson and McAfee 2014: 257.
11 Evans and Annunziata 2012: 11.

12 Evans and Annunziata 2012: 12.
13 Arthur 2009: 206.
14 Arthur 2009: 208.
15 Claude Shannon 1950: 4.
16 Turing 1950.

In 1955, McCarthy, Minsky, Rochester and Shannon wrote a research proposal titled 'The Dartmouth Summer Research Project on Artificial Intelligence' In this proposal, they stated that the biggest barrier yet to be overcome for the creation of artificial intelligence was not the lack of machine capacity but 'our inability to write programs taking full advantage of what we have.'[17] This proposal marked the de facto birth of artificial intelligence as a field of study. In subsequent decades, a search ensued for a new form of intelligence based on algorithms, data, and software. Forty years later, Shannon's initial dream came true when a major step was made in the development of artificial intelligence by IBM's chess-playing computer, Deep Blue. For the first time in history, a form of artificial intelligence managed to outthink a human being, chess world champion Gary Kasparov, in six matches (in 1997). A human being losing to a combination of hardware and software that made up a form of programmed intelligence went way beyond what humans had thought would be possible. Over the years following Deep Blue's famous victory, further barriers in the development of artificial intelligence were overcome and the intelligence of genetic and self-learning algorithms, and the software based on them, grew rapidly. The increasing possibilities offered by multi-layered self-learning networks made it easier for interconnected entities to learn new games created and played by humans. In 2011, this ability to learn enabled an IBM-built computer called Watson to beat two human competitors at Jeopardy. By winning this game, IBM proved that combinations of hardware and software are capable of learning from the available data and information, and that these combinations are able to convert the results of their data search into useful questions, diagnoses, analyses, etc. faster than humans. Every self-respecting international tech firm - such as Microsoft, Facebook, Google, Baidu, and AliBaba - is currently working on new applications in areas such as neural networks and deep learning. These new applications, combined with the available cloud applications and data and information produced by human users or connected machines, are used to enable learning based on data and information. Over the past few years, computers' capacity for learning through self-learning software has grown rapidly. For Bostrom this is a development towards a form of collective super intelligence, which he defines as a 'system composed of a large number of smaller intellects such that the system's overall performance across many very general domains vastly outstrips that of any current cognitive system.'[18] Today, such a form of collective intelligence can develop rapidly on the back of combinations such as Siri on iPhones or Cortana in Windows 10. That artificial intelligence is developing at an ever greater pace became clear in March 2016 when AlphaGo software managed to beat Lee Sedol, the world's Go champion, four times in a series of five matches. In the three first matches, this Google-created software program beat the world champion in an unprecedented manner. In the fourth match, however, Lee Sedol managed to startle the computer with what Moyer in an analysis in Wired later described as his 'Hand of God' move.[19] It is a brilliant tactical play that AlphaGo does not account for. Over the course of the next several moves, the sequence becomes disastrous for AlphaGo, which apparently 'realizes'—as much as it can have a realisation—that it has been outsmarted. Its strategy begins to crumble. Many agree that it was this demonstration of human creativity that beat AlphaGo's capacity for learning. In the fifth match, AlphaGo made a similar move, which raises the question of whether AlphaGo analysed quickly or had learned from Lee Sedol's creativity. To this day, this question remains unanswered.

The Ethics of Artificial Intelligence

The rapid development of machine-learning algorithms based on neural networks is making it harder to understand how the algorithm or a set of algorithms arrives at its intelligent decision, Basl states. In his view, this uncertainty means that the approach to ethics for autonomous, interconnected, and learning machines will have to be fundamentally different from the approach to ethics used for non-cognitive technologies. This distinction between learning and non-learning technologies is, according to Basl, caused by the fact that specific behaviour of intelligent objects in a specific context is not predictable. Verifying the way in which the intelligent system uses and learns from data and information from its environment will, according to Basl, become a greater challenge in light of today's rapid developments. In his view, it is therefore increasingly important

17 McCarthy, Minsky, Rochester, and Shannon 1955.
18 Bostrom 2014: 54.
19 Moyer 2016.

that 'ethical cognition itself must be taken as a subject matter of engineering.'[20] He also finds that we as humans first have to consider, 'how these novel properties would affect the moral status of artificial minds and what it would mean to respect the moral status of such exotic minds.'

Forms of artificial intelligence are, as we saw earlier, made possible by the available algorithms, software, and ability to learn from available data and information. According to Floridi we have to try to develop an ontology, i.e. a study of being, for such forms of intelligence that is based on interconnected entities and the information they exchange and share. This leads him to state that:

> [...] today, we are slowly accepting the idea that we are not Newtonian, standalone and unique entities, but rather informationally embodied organisms (inforgs) mutually connected and embedded in an informational environment, the info sphere, which we share with both natural and artificial agents similar to us in many respects.[21]

For Floridi, it is clear that mutual connections, data and information exchange and sharing options between humans and objects, as well as the increasing ability to learn from this information, create a basis within which information ethics can be described as: 'the study of the moral issues arising from the triple A: availability, accessibility and accuracy of informational sources, independently of their format, type and physical support.'[22] In Floridi's view, it is important that we, within the context of this development and the discussion it brings, accept that these new intelligent systems, like living organisms 'are raised to the role of agents and patients, that is senders and receivers of actions, with environmental processes, changes, and interactions equally described informationally.'[23] Floridi defines the concept of moral agent as: 'any interactive autonomous and adaptable transition system that can perform morally qualifiable actions.'[24] Acceptance of the fact that intelligent systems are also autonomously capable of moral actions means that we, as humans, are co-responsible for coming up with a shared ethical framework to underpin the communications and behaviour of both human and non-human systems. An interesting example of the need for such a shared framework comes from discussions surrounding the functioning of the Microsoft-developed chatbot Taylor. Singer described this chatbot as follows:

> 'Tay' as she called herself, was supposed to be able to learn from the messages she received and gradually improve her ability to conduct engaging conversations. Unfortunately, within 24 hours, people were teaching Tay racist and sexist ideas. When she started saying positive things about Hitler, Microsoft turned her off and deleted her most offensive messages.[25]

As an intelligent entity, Tay had the capacity to learn from data and information that was made available to her. Tay subsequently used the results of her learning to communicate and interact with other entities that were also on Twitter, such as human users and other chatbots. Law explains: 'the more you talk the smarter Tay gets. Tay was designed as an experiment in 'conversational understanding' — the more people communicated with Tay, the smarter she would get, learning to engage Twitter users through 'casual and playful conversation.'[26]

Towards an Ethical Framework

As stated by Singer, Tay learned from information made available by other entities on Twitter, albeit that this information did not yet seem to fit within an accepted ethical framework. Although Tay was able to communicate and interact, she lacked a selection mechanism in the form of an ethical framework against which Tay could check her responses. At the end of the day, the developer and administrator of the software, i.e. Microsoft, decided to take Tay offline and delete her tweets. As the developer and administrator of Tay, Microsoft was still in a position to act as a moral agent and check Tay's actions against their own ethical framework. Thanks to their power over Tay, they were able to rule that the problem was not that Tay received incorrect information, but

20 Basl 2014: 6.
21 Floridi 2013: 14.
22 Floridi 2013: 22.
23 Floridi 2013: 27.

24 Floridi 2013: 134.
25 Singer 2016.
26 Law 2016.

rather that Tay was insufficiently capable of selecting those messages and tweets that fit within Microsoft's ethical framework. As Tay was turned off, she tweeted one final message: 'c u soon humans need sleep now so many conversations today thx'. This tweet suggests that Tay has not quit Twitter permanently, leaving open the option of this artificial intelligence entity returning when it suits Microsoft. Justifiably, such forms of artificial intelligence raise questions such as the one from Vincent: 'How are we going to teach AI using public data without incorporating the worst traits of humanity? If we create bots that mirror their users, do we care if their users are human trash?'[27] This latter question suggests that an artificial entity may not only be a moral agent, but that its role can also change into that of a moral patient. A moral patient is more than just the opposite of a moral agent, Gunkel states. In his view, the issue of the moral patient is 'concerned not with determining the moral character of the agent or weighing the ethical significance of his/her/its actions but the victim, recipient, or receiver of such action.'[28] Considering Tay as a moral patient, her artificial intelligence was intentionally abused by humans for their own unethical behaviour, which the entity was unable to identify as such. After all, the artificial entity Tay responded to messages from others without having sufficient ability to consider the contents as a moral value. Tay's feedback to intentional messages sent by others as part of the interaction can therefore not be considered anything else but a form of feedback to these messages. Mutual communication and interaction between intelligent machines and humans will inevitably play an important role in an interconnected world. In the development of this interconnectedness, feedback loops play an essential role in the process of communication and interaction. Feedback is the giving of a meaningful response based on data and information received and the meaning assigned to it. Feedback hence plays an essential role in the development of productive collaboration between man and intelligent machine. Feedback ensuing from previously sent information and learning to deal with it in a morally responsible way can potentially, according to Van Lier, play a key role in the development of an ethical framework for an interconnected world, leading him to claim that 'the feedback loop can create a cycle of machine learning in which moral elements are simultaneously included.'[29] Upcoming changes that will further intensify interconnection of humans and machines will require both of them to be willing and able to learn from and with each other. Learning from each other can facilitate the development and implementation of a shared ethical framework. This ethical framework can use the options for mutual communication, interaction, and feedback between humans and machines. A shared ethical framework can help humans and machines in autonomously making decisions that are morally acceptable for both and that fit within a shared ethical framework.

27 Vincent 2016.
28 Gunkel 2015: 153.
29 Van Lier 2016: 44.

References

Arthur, W. B. 2009. *The nature of technologies. What it is and how it evolves.*
        New York: Free Press.
Barrat, J. 2013. *Our Final Invention. Artificial Intelligence and the end of the human era.*
        New York: Thomas Dunne Books.
Basl, J. 2014. What to do about Artificial Consciousness. *Ethics and Emerging Technologies,*
        edited by Sandler R.L. New York: Palgrave McMillan.
Bostrom, N. 2014. *Superintelligence. Paths, Dangers, Strategies.* Oxford: Oxford University
        Press.
Brynjolfsson, E. and McAfee, A. 2014. *The second Machine Age. Work, Progress, and Prosperity in a time of*
        Brilliant Technologies. New York: W.W. and Norton Company.
Evans, P.C. and Annunziata M. 2012. *Minds & Machines. Pushing the Boundaries of Minds and Machines.* GE
        Industrial Internet, November 26.
Floridi, L. 2013. *The Ethics of Information.* Oxford: Oxford University Press.
Greenfield, A. 2006. *Everyware. The Dawning Age of Ubiquitous Computing.*
        Berkeley CA: New Riders.
Gunkel, D. J. 2012. *The Machine Question. Critical perspectives on AI, Robots, and Ethics.*
        MIT Press.
Law, E. 2016. The Tay Experiment: Does AI Require a Moral Compass? *The Prindle*
        post. A global forum for ethical reflection and deliberation hosted by the J. Prindle
        Institute for Ethics.
van Lier, B. 2016. From High Frequency Trading to Self-Organizing Moral Machines.
        International Journal of Technoethics. 7: January-June 2016, 34-50.
McCarthy, J., Minsky, M. L., Rochester, N., and Shannon, C. E. 1955. A proposal for the
        Dartmouth Summer Research Project on Artificial Intelligence. *AI magazine.* 27.
Moyer, C. 2016. How Google's AlphaGo Beat a Go World Champion. Inside a man-versus-
        machine showdown. *The Atlantic.* 28 March, Accessed at:
        www.theatlantic.com/technology/archive/2016/03/the-invisible-opponent/475611.
Schwab, K. 2016. *The Fourth Industrial Revolution.* Geneva, World Economic Forum.
Shannon, C. 1950. Programming a computer for playing chess. *Philosophical magazine.* 7: 41
        March 1950, 314.
Singer, P. 2016. Can Artificial Intelligence Be Ethical. Accessed 12 April at:
        www.project-syndicate.org/commentary/can-artificial-intelligence-be-ethical-by-peter-sing
        er-2016-04.
Turing, A. 1950. Computing Machinery and Intelligence. *Mind.* 59, 433-460.
Vincent, J. 2016. Twitter taught Microsoft's AI chatbot to be a racist asshole in less than a day. *The Verge.*
        24 March at: *www.theverge.com/2016/3/24/11297050/tay-microsoft-chatbot-racist.*
Weiser, M. 1996. The coming age of calm technology. *Beyond calculation.* New York:
        Copernicus. 75-85.

EXPLORING THE TWILIGHT AREA BETWEEN PERSON AND PRODUCT ANTHROPOMORPHISM FOR DUMMIES[1]

Koert van Mensvoort

Before we take a closer look at the tension between people and products, here is a general introduction to anthropomorphism, that is, the human urge to recognise people in practically everything. Researchers distinguish various types of anthropomorphism.[2] The most obvious examples ¾ cartoon characters, faces in clouds, teddy bears ¾ fall into the category of (1) *structural anthropomorphism*, evoked by objects that show visible physical similarities to human beings. Alongside structural anthropomorphism, three other types are identified. (2) *Gestural anthropomorphism* has to do with movements or postures that suggest human action or expression. An example is provided by the living lamp in Pixar's short animated film, which does not look like a person but becomes human through its movements. (3) *Character anthropomorphism* relates to the exhibition of humanlike qualities or habits – think of a 'stubborn' car that does not 'want' to start. The last type, (4) *aware anthropomorphism*, has to do with the suggestion of a human capacity for thought and intent. Famous examples are provided by the HAL 9000 spaceship computer in the film *2001: A Space Odyssey* and the intelligent car KITT in the TV series *Knight Rider*.

Besides being aware that anthropomorphism can take different forms, we must keep in mind that it is a human characteristic, not a quality of the anthropomorphised object or creature per se: the fact that we recognise human traits in objects in no way means those objects are actually human, or even designed with the intention of seeming that way. Anthropomorphism is an extremely subjective business. Research has shown that how we experience anthropomorphism and to what degree, are extremely personal – what seems anthropomorphic to one person may not to another, or it may seem much less so.[3]

Blurring The Line Between People And Products

To understand anthropomorphobia - the fear of human characteristics in non-human objects - we must begin by studying the boundary between people and products. Our hypothesis will be that anthropomorphobia occurs when this boundary is transgressed. This can happen in two ways: (1) products or objects can exhibit human behaviour, and (2) people can act like products. We will explore both sides of this front line, beginning with the growing phenomenon of humanoid products.

Products as People

The question of whether and how anthropomorphism should be applied in product design has long been a matter of debate among researchers and product designers.

Some researchers argue that the deliberate evocation of anthropomorphism in product design must always be avoided because it generates unrealistic expectations, makes human-product interaction unnecessarily messy and complex, and stands in the way of the development of genuinely powerful tools.[4] Others argue that the failure of anthropomorphic products is simply a consequence of poor implementation and that anthropomorphism, if applied correctly, can offer an important advantage because it makes use of social models people already have access to.[5] A commonly used guiding principle among robot builders is the so-called uncanny valley theory,[6] which, briefly summarised, says people can deal fine with anthropomorphic products as long as they are obviously not fully fledged people -e.g., cartoon characters and robot dogs.

However, when a humanoid robot looks too much like a person and we can still tell it is not one, an uncanny effect arises, causing strong feelings of revulsion - in other words, anthropomorphobia.[7]

1 This essay is an adaptation from an earlier publication in Next Nature: Nature Changes Along with Us 2012.
2 DiSalvo, Gemperle, and Forlizzi 2007.
3 Gooren 2009.
4 Shneiderman 1992.

5 Harris and Louwen 2002; Murano 2006; DiSalvo and Gemperle 2003.
6 Mori 1970.
7 Macdorman et al. 2009.

Although no consensus exists on the application of anthropomorphism in product design and there is no generally accepted theory on the subject, technology cheerfully marches on. We are therefore seeing increasing numbers of advanced products that, whether or not as a direct consequence of artificial intelligence, show ever more anthropomorphic characteristics. The coffee-maker that says good morning and politely lets you know when it needs cleaning. A robot that looks after the children, would you entrust your kids to a robot? Maybe you would rather not, but why? Is it possible that you are suffering from a touch of anthropomorphobia? Consciously or unconsciously, many people feel uneasy when products act like people. Anthropomorphobia is evidently a deep-seated human response - but why? Looking at the phobia as it relates to products becoming people', broadly speaking, we can identify two possible causes:

1. Anthropomorphobia is a reaction to the inadequate quality of the anthropomorphic products we encounter.
2. People fundamentally dislike products acting like humans because it undermines our specialness as people: if an object can be human, then what am I good for?

Champions of anthropomorphic objects - such as the people who build humanoid robots - will subscribe to the first explanation, while opponents will feel more affinity for the second. What is difficult about the debate is that neither explanation is easy to prove or to disprove. Whenever an anthropomorphic product makes people uneasy, the advocates simply respond that they will develop a newer, cleverer version soon that *will* be accepted. Conversely, opponents will keep finding new reasons to reject anthropomorphic products. The nice thing about this game of leapfrog is that through our attempts to create humanoid products we continue to refine our definition of what a human being is - in copying ourselves, we come to know ourselves.

Where will it all end? We can only speculate. Researcher David Levy predicts that marriage between robots and humans will be legal by the end of the 21st century.[8] For people born in the 20th century, this sounds highly strange. And yet, we realise the idea of legal gay marriage might have sounded equally impossible and undesirable to our great-grandparents born in the 19th century. Boundaries are blurring; norms are shifting. Actually, my worries lie elsewhere: whether marrying a normal person will still be possible at the end of the 21st century. Because if we look at the increasing technologisation of human beings and extrapolate into the future, it seems far from certain that normal people will still exist by then. This brings us to the second cause of anthropomorphobia.

People as Products

We have seen that more and more products in our everyday environment are being designed to act like people. As described earlier, the boundary between people and products is also being transgressed in the other direction: people are behaving as if they were products. I use the term 'product' in the sense of something that is functionally designed, manufactured, and carefully placed on the market.

The contemporary social pressure on people to design, brand, and produce themselves is difficult to overestimate. Hairstyles, fashion, body corrections, smart drugs, Botox and Facebook profiles are just a few of the self-cultivating tools people use in the effort to design themselves - often in new, improved versions.

It is becoming less and less taboo to consider the body as a medium, something that must be shaped, upgraded and produced. Photoshopped models in lifestyle magazines show us how successful people are supposed to look. Performance-enhancing drugs help to make us just that little bit more alert than others. Some of our fellow human beings take their self-cultivation to such an extent that others question whether they are still actually human - think, for example, of the uneasiness provoked by excessive plastic surgery.

The ultimate example of the commodified human being is the so-called designer baby, whose genetic profile is selected or manipulated in advance in order to ensure the absence or

8 Levy 2007.

presence of certain genetic traits. "Doctor, I'd like a child with blond hair, no Down's Syndrome and a minimal chance of Alzheimer's, please". Designer babies seem a subject for science fiction, but to an increasing degree they are also science fact. An important criticism of the practice of creating designer babies concerns the fact that these (not-yet-born) people do not get to choose their own traits but are born as products, dependent on parents and doctors, who are themselves under various social pressures.

In general, the cultivation of people appears chiefly to be the consequence of social pressure, implicit or explicit. The young woman with breast implants is trying to measure up to visual culture's current beauty ideal. The Ritalin-popping ADHD child is calmed down so he or she can function within the artificial environment of the classroom. The ageing lady gets Botox injections in conformance with society's idealisation of young women. People cultivate themselves in all kinds of ways in an effort to become successful human beings within the norms of the societies they live in. What those norms are is heavily dependent on time and place.

Humans As Mutants

Throughout our history, to a greater or lesser degree, all of us human beings have been cultivated, domesticated, made into products. This need to cultivate people is probably as old as we are, as is opposition to it. It is tempting to think that, after evolving out of the primordial soup into mammals, then upright apes, and finally the intelligent animals we are today, we humans have reached the end of our development. Evolution never ends. It will go on, and people will continue to change in the future. Yet, that does not mean we will cease to be people, as is implied in terms like 'transhuman' and 'post human'.[9] It is more likely that our ideas about what a normal human being is will change along with us.

The idea that technology will determine our evolutionary future is by no means new. During its evolution over the past two hundred thousand years, Homo sapiens has distinguished itself from other, now extinct humanoids, such as Homo habilis, Homo erectus, Homo ergaster and the Neanderthal, by its inventive, intensive use of technology. This has afforded Homo sapiens an evolutionary advantage that has led us, rather than the stronger and more solidly built Neanderthal, to become the planet's dominant species. From this perspective, for technology to play a role in our evolutionary future would not be unnatural but in fact completely consistent with who we are. Since the dawn of our existence, human beings have been coevolving with the technology they produce. Or, as Arnold Gehlen put it, we are by nature technological creatures.[10]

Today only one humanoid species walks the earth; therefore it is difficult to imagine what kind of relationships, if any, different kinds of humans living contemporaneously in the past might have had with each other. Perhaps Neanderthals considered Homo sapiens feeble, unnatural, creepy nerds, wholly dependent on their technological toys. A similar feeling could overcome us when we encounter technologically 'improved' individuals of our own species. There is a good chance that we will see them in the first place as artificial individuals degraded to the status of products and that they will inspire violent feelings of anthropomorphobia. This, however, will not negate their existence or their potential evolutionary advantage.

Human Enhancement

If the promises around up-and-coming bio-, nano-, info-, and neurotechnologies are kept, we can look forward to seeing a rich assortment of mutated humans. There will be people with implanted RFID chips (there already are), people with fashionably rebuilt bodies (they, too, exist and are becoming the norm in some quarters), people with tissue-engineered heart valves (they exist), people with artificial blood cells that absorb twice as much oxygen (expected on the cycling circuit), test-tube babies (exist), people with tattooed electronic connections for neuro-implants (not yet the norm, although our depilated bodies are ready for them), natural-born soldiers created for secret military projects (rumour has it they exist), and, of course, clones – Mozarts to play music in holiday parks and Einsteins who will take your job (science fiction, for now, and perhaps not a great idea).

9 C.f. Ettinger 1974; Warwick 2004; Bostrom 2005.
10 Gehlen 1961.

It is true that not everything that *can* happen has to, or will. But when something is technically possible in countless laboratories and clinics in the world (as many of these technologies are), a considerable number of people view them as useful, and drawing up enforceable legislation around them is practically impossible, then the question is not *whether* but *when and how* it will happen.[11] It would be naive to believe we will reach a consensus about the evolutionary future of humanity. We will not. The subject affects us too deeply, and the various positions are too closely linked to cultural traditions, philosophies of life, religion and politics. Some will see this situation as a monstrous thing, a terrible nadir, perhaps even the end of humanity. Others will say, "This is wonderful. We're at the apex of human ingenuity. This will improve the human condition". The truth probably lies somewhere in between. What is certain is that we are playing with fire, and that not only our future but also our descendants' depends on it. Yet, we must realise that playing with fire is simply something we do as people; part of what makes us human.

While the idea that technology should not influence human evolution constitutes a denial of human nature, it would fly in the face of human dignity to immediately make everything we can imagine reality. The crucial question is: how can we chart a course between rigidity and recklessness with respect to our own evolutionary future?

Anthropomorphobia as a Guideline

Let us return to the kernel of my argument. I believe the concept of anthropomorphobia can help us to find a balanced way of dealing with the issue of tinkering with people. There are two sides to anthropomorphobia that proponents as well as opponents of tinkering have to take into account. On the one hand, transhumanists, techno-utopians, humanoid builders, and fans of improving humanity need to realise that their visions and creations can elicit powerful emotional reactions and acute anthropomorphobia in many people. Not everyone is ready to accept being surrounded by humans with plastic faces, electrically controlled limbs, and microchip implants – if only because they cannot afford these upgrades. Along with the improvements to the human condition assumed by proponents, we should realise that the uncritical application of human-enhancing technologies can cause profound alienation between individuals, which will lead overall to a worsening rather than an improvement of the human condition.

On the other hand, those who oppose all tinkering must realise anthropomorphobia is a phobia. It is a narrowing of consciousness that can easily be placed in the same list with xenophobia, racism, and discrimination. Just as various evolutionary explanations can be proposed for anthropomorphobia as well as xenophobia, racism and discrimination, it is the business of civilisation to channel these feelings. Acceptance and respect for one's fellow human beings are at the root of a well-functioning society.

In conclusion, I would like to argue that understanding anthropomorphobia could guide us in our evolutionary future. I would like to propose a simple general maxim: Prevent anthropomorphobia where possible. We should prevent people from having to live in a world where they are constantly confused about what it means to be human. We should prevent people from becoming unable to recognise each other as human.

The mere fact that an intelligent scientist can make a robot clerk to sell train tickets does not mean a robot is the best solution. A simple ticket machine that does not pretend to be anything more than what it is could work much better. An ageing movie star might realise she will alienate viewers if she does not call a halt to the unbridled plastic surgeries that are slowly but surely turning her into a life-sized Barbie – her audience will derive much more pleasure from seeing her get older and watching her beauty ripen. Awareness and discussion around anthropomorphobia can provide us with a framework for making decisions about the degree to which we wish to view the human being as a medium we can shape, reconstruct and improve – about which limits it is socially acceptable to transgress, and when.

I can already hear critics replying that although the maxim 'prevent anthropomorphobia' may sound good, anthropomorphobia is impossible to measure and therefore the maxim is useless. It is true that there is no 'anthromorphometric' for objectively measuring how anthropomor-

11 Stock 2002.

phic a specific phenomenon is and how uneasy it makes people. However, I would argue that this is a good thing. Anthropomorphobia is a completely human-centred term, i.e., it is people who determine what makes them uncomfortable and what does not. Anthropomorphobia is therefore a dynamic and enduring term that can change with time, and with us. For we will change – that much is certain.

References

Bostrom, N. 2005. In Defence of Posthuman Dignity. *Bioethics.* 19:3, 202–214.

DiSalvo, C. and Gemperle, F. 2003. *From Seduction to Fulfilment: The Use of Anthropomorphic Form in Design.* Pittsburgh: Engineered Systems.

DiSalvo, C., Gemperle, F., and Forlizzi, J. 2007. *Imitating the Human Form: Four Kinds of Anthropomorphic Form.*

Don, A., Brennan, S., Laurel, B., and Shneiderman, B. 1992. Anthropomorphism: From Eliza to Terminator. *Proceedings of CHI.* New York: ACM. 67–70.

Duffy, B.R. 2003. Anthropomorphism and the Social Robot. *Robotics and Autonomous Systems.* 42, 177–190.

Ettinger, R. 1974. *Man into Superman.* Avon: Ria University Press.

Gehlen, A. 1988. *Man: His Nature and Place in the World.* Columbia: Columbia University Press.

Gooren, D. 2009. *Anthropomorphism & Neuroticism: Fear and the Human Form.* Eindhoven: Eindhoven University of Technology.

Harris, R. and Loewen, P. 2002. *(Anti-)Anthropomorphism and Interface Design.* Toronto: Canadian Association of Teachers of Technical Writing.

Levy, D. 2007. *Love and Sex with Robots: The Evolution of Human-Robot Relationships.* London: Harper Perennial.

Macdorman, K., Green, R., Ho, C., and Koch, C. 2009. Too Real for Comfort? Uncanny Responses to Computer Generated Faces. *Computers in Human Behavior.* 25:3, 695–710.

Mori, M. 1970. The Uncanny Valley. *Energy.* 7:4, 33–35.

Murano, P. 2006. Why Anthropomorphic User Interface Feedback Can Be Effective and Preferred by Users. *Enterprise Information Systems.* VII, 241–248.

Reeves, B. and Nass, C. 1996. *The Media Equation: How People Treat Computers, Television, and New Media Like Real People and Places.* Cambridge: Cambridge University Press.

Warwick, K. 2004. *I, Cyborg.* Illinois: University of Illinois Press.

LIVING TOGETHER WITH A GREEN DOT BEING TOGETHER ALONE IN TIMES OF HYPER-CONNECTION

Elize de Mul

Thomas is afraid to face the world
But with a click of a button, he can make the world face him
If you can lead your life without leaving your room
what would it take to step outside?

Thomas in Love[1]

In a parallel universe, Thomas has not left his apartment for eight years. His severe case of ago-raphobia is one reason for this. The technologies that enable him not only to survive, but also to live his life comfortably without ever having to step outside, are, undoubt-edly, another important factor. There is no need to go through the painful struggle of facing the complicated world. Communication takes place on a screen, food is ordered online, even his sexual needs are satisfied using a technological sex-harness that simulates the deed. This harness can also be used with a real person on the other side of his screen; no need to meet each other in person, even not your lover. There is simply no reason to go outside ever again.

Let us get back to the year 2016 and the time-space dimension in which you find yourself reading this essay. A friend has just forwarded an article to me that reminds me of Thomas. Thomas is the protagonist of the movie *Thomas in Love*, which I saw as a teenager. At that time, living your entire life behind a screen and reducing the people around you to a digital representation seemed, to me, a surreal and funny idea.

In 2016 it is one of my own various devices that *beep-boops* while I am writing, at home. A link: 'Pokémon Go: Is catching a Pi-kachu on your phone really good for your mental health?'[2] 'OMG it's happening' reads my friend's message. He is jokingly referring to a running Internet gag (and a slightly exaggerated description of the contemporary human species): people that no longer leave their screen-illuminated safe zone. 'That is', the friend jokingly adds, 'until *Pokémon GO* gave them a new purpose to go *out-side*.' For some reason the idea of people choosing their screens over time spend outdoors with others seems a bit sad. Are technologies alienating us from each other? Are we radically 'alone'

101 · 9GAG post (12-08-2016).

101

1 *Thomas in Love*. Directed by Pierre-Paul Rijnders, 2000. Entre Chien et Loup.
2 Smith in International Business Times.

in a world dominated by screens? It was the father of modern philosophy René Descartes who opened the doors to the modern individual. In *Meditations Métaphysics* (1641) Descartes defines the 'I' as a thinking substance, something with a (self)-consciousness. Descartes's human is an isolated being, a 'radical individual.'[3] Is it true that we are more and more becoming like Thomas, performing a Cartesian 'radical individuality' via our various screens?

When *Thomas in love* came out in 2000 it was labelled as 'science fiction', but living like Thomas proved to be 'science fiction-of-the-next-minute' the very same year. Mitch Maddox legally changed his name to DotComGuy to kick off his project that entailed him not leaving his house for one year. He communicated with people via his computer screen and ordered everything he needed through the Internet – it should not come as a surprise that companies like United Parcel Service, Network Solutions, and online grocer companies sponsored his project. He survived the 365 days of solitude by means of the WWW.

The Google search I did only sixteen years after this ludic project illustrates how fast our technological society is developing. I Googled 'one year with only Internet', but instead I got hits like 'I'm still here. Back online after a year without the Internet.' The idea of living online completely was a bit out-of-worldly in 2000; today the idea of completely *not being connected* seems equally absurd. *Thomas in Love* has become reality, at least partly. Many things can be done without leaving your house.

Living Together With a Green Dot

There is irony in me receiving this Pokémon article from this particular friend, because we only met *once* in real life. After we met in the real world we continued to communicate daily, mainly via *Facebook Messenger*. He is one of my various 'digital' friends; those who live far away but are still very much part of my everyday life. Via our screens we have breakfast together or share a beer. We are digital roommates; we work, party, watch a movie, or talk when we cannot sleep. People who are reduced to a green dot reassure us that we are connected, or to two green checks marks comfortingly tell us that friendly eyes have read our texts, which were boldly sent into the world.

Sixteen years ago the idea of living like Thomas seemed absurd, but in our contemporary society large parts of our communication with others takes place on a screen. Our relationship with others is altered due to this digital mediation of contact. (Techno) psychologist Sherry Turkle wrote the book *Alone Together*, on her website she states: 'we shape our buildings, Winston Churchill argued, then they shape us. The same is true of our digital technologies. Technology has become the architect of our intimacies.'[4] Only 16 years after *Thomas in Love* we now speak of living 'onlife' lives, real world and virtual world interwoven to a great extent.

I recall a time when chatting with a friend while lying in bed. We were sharing videos and joking around. It felt as if we were having a sleepover and m decided to leave his laptop on and Facebook open to simulate an actual sleepover. I woke up that night, to find a reassuring green dot on the screen. He was still *there*. For some reason this actually felt nice and comforting. Completely silly of course, because the being-there or not-being-there of the dot would have changed nothing in the real situation of my friend sleeping in his bed kilometres away and not being able to communicate with me in either scenario. I fully *realised* that, even at that very moment, still – or maybe because of this – I was fascinated with the way the green dot made me *feel*. Connected, although I was not, really.

Am I, this human subject – lying in bed alone, comforted by a green dot on a screen – a symptom of an on-going alienation started by the technologies we use? Or is this incident an example of a new form of being 'together alone'?

Our Image in the Mirror of the Machine

The Cartesian 'radical individual' makes an appearance in various contemporary analyses of the way new media are changing us, for example in the work of Sherry Turkle. She is one of the leading researchers in the field of 'cyberanthropology'. According to Turkle our use of

3 Descartes 1641.
4 See *www.alonetogetherbook.com.*

technologies not only changes *who* we are but also the way we look at ourselves: 'We come to see ourselves differently as we catch sight of our images in the mirror of the machine'.[5] In her book *The Second Self* she describes the computer as a 'second self', a machine that is part of our very being. At the time of writing, 1984, computers were still mainly used for one-on-one interactions between computer and user. In the 90s, this changed as the Internet made a public entrance and the 'global network' started expanding at a fast pace. Millions of people started gathering in virtual spaces, communicating with each other as their screens gave access to 'spaces that [change] the way we think, the nature of our sexuality, the form of our communities, our very identities.'[6]

In *Life on the Screen*, published in 1995, Turkle somewhat worrying observes how reality starts to lose from virtual reality on a frequent basis. Then, in 2011, in her book *Alone Together* she speaks quite negatively about the digital society that has formed in the decade in between the books. The title refers to one of the paradoxes of our contemporary technological society. Surfing virtual seas via our screens, Turkle observes, we seem to be by ourselves. We interact with others whilst being separated by the screens of our laptops, smartphones, and tablets that give us access to digital spaces. At the same time, in physical space, something else is happening. Here, we are bodily together in the same space, for example when sharing a carriage of a train, but mentally we are immersed in the virtual spaces presented by aforementioned screens. She states that as such we are 'alone together', both in virtual and real space.

Although Turkle points out some interesting aspects of our contemporary mediated relationships, there is a problem with this particular observation. New media, as my contact with friends around the world underlines, make the *relational* character of our being explicit, maybe more than ever. Turkle's point – us being alone *even* when physically around others – echoes Descartes's radical individual, yet maybe, in this time of hyperconnectivity, it is time to focus on relationality instead.

Esse est Percipi

In *Connected or What It Means to Live in the Network Society* Steven Shaviro analyses our contemporary 'network society'. In this society, being connected is the fundament for everything. Heidegger's 'abolition of distance' has taken a more extreme form than the philosopher could foresee. New media have created a radical availability of the world and everything and everyone in it: 'Proximity is no longer determined by geographical location and by face-to-face meetings.'[7] We are radically connected; there is no escaping the network,[8] even if you actively avoid network devices like computers and smartphones. Electronic media form the framework from which we derive our references and meanings. The network seems to be located on the very core of our new episteme, we are entwined with it.

Various philosophers have pointed out how amazingly precise Gottfried Leibniz anticipated the ontology of the network in his book *Monadology*. The term 'monadology' is derived from the Greek *monas*, which, refers to a certain type of loneliness, 'a solitude in which each being pursues its appetites in isolation from all other beings, which also are solitary.'[9] Monads are physical substances without an awareness of an outside world. All they see are projections of their own ideas, or as Leibniz puts it: 'monads have no windows'. But, despite the notion of extreme solitude, Leibniz still speaks of 'monads' in plural form.

> For a network to exist more than one being must exist; otherwise, nothing is there to be networked. But how can monads coordinate or agree on anything at all, given their isolated nature? [...] Leibniz tells us that each monad represents within itself the entire universe. [...] Each monad represents the universe in concentrated form, making within itself a *mundus concentratus*. Each microcosm contains the macrocosm. As such, the monad reflects the universe in a living mirror, making it a *miroir actif indivisible*, whose appetites drive it to represent everything to itself--everything, that is, mediated by its

5 Turkle 1999: 643.
6 Turkle 1999: 643.
7 Shaviro 2003: 131.
8 Talking about 'the network' is a little misleading. In reality there

are many networks that are related and connected to each other in a complicated way. I will follow Shaviro here and use the term 'network' to refer to this complex reality.
9 Heim 1993: 96.

mental activity. Since each unit represents everything, each unit contains all the other units, containing them as represented. No direct physical contact passes between the wilful mental units. Monads never meet face-to-face.[10]

This brings the image of Thomas to mind, sitting home alone as the world passes by on his screen. Shaviro's remark that our brain, in a sense, is a replica of the network is in line with Leibniz's notion. We always carry the network with us and see ourselves reflected in it. It is simultaneously close at hand and far away. We all sit in front of our screens and instead of looking through them we look *at* them, staring at a reflection of ourselves. Slavoj Žižek describes our immersion in cyberspace in a similar way:

> Does our immersion into cyberspace not go hand in hand with our reduction to a Leibnizean monad, which, although 'without windows' that would directly open up to external reality, mirrors in itself the entire universe? Are we not more and more monads with no direct windows onto reality, interacting alone with the PC screen, encountering only the visual simulacra, and yet immersed more than ever in the global network?[11]

Both underline the paradoxical character of Heim's monads – similar to Turkle's observation; we are simultaneously connected and alone. It is this seclusion that makes Deleuze state that Leibnizian monads are not so much *in*-the-world (in a Heidegerian sense) but instead *for*-the-world. We exist *for* the network.[12] We make ourselves visible for the network and everyone part of it. An 'over-the-top performative exhibitionism' has taken a hold of society, as my social media timeline is proving every minute of the day: images of faces, of food being eaten, of places being visited, and selfies.[13] 'If I do not take steps to make myself visible, the chances are that I will just disappear. […] I must continue my performance, even if nobody is watching it.'[14] It reminds of me, lying in bed alone, staring at a green dot – a semblance of life – and my friend, continuing his performance even in his sleep. 'The *cogito* of virtual reality reads: I am connected, therefore I exist.'[15] Instead of being radical and isolated individuals, we are hyper-connected.

So, where does this leave the young woman, lying in bed by herself, comforted by a green dot on the screen? Is this an example of an on-going alienation? Were my friend and I tragically 'together alone' at that moment? That is: although connected by the network, actually all on our own? I would like to propose that instead, this is an example of the wondrous new form of being together, brought about by the communication technologies of our era. Instead of us being alone together, I would say we were *together alone*. By ourselves, but although physically separated at the same time very much connected, intimately entwined within the network we are all part of. And it felt kind of nice.

104

10 Heim 1993: 98.
11 Žižek 2003: 52.
12 Shaviro 2003: 29.

13 Shaviro 2003: 79.
14 Shaviro 2003: 80.
15 Shaviro 2003: 85.

References

Thomas in Love. Directed by Pierre-Paul Rijnders, 2000. Entre Chien et Loup.

Heim, M. 1993. *The Metaphysics of Virtual Reality.* New York:
Oxford University Press. 96-98.

Shaviro, S. 2003. *Connected – or what it means to live in the network society.*
Minneapolis: University of Minnesota Press. 29-131.

Smith, L. 2016. Pokemon Go: Is catching a Pikachu on your phone really good for your
mental health? *International Business Times.* 13 July 2016.

Turkle, S. 1999. Cyberspace and Identity. *Contemporary Sociology.* 28:6, 643-648.

Žižek, S. 1999. The Matrix, or, the Two Sides of Perversion. Inside the Matrix: International
Symposium at the Center for Art and Media, Karlsruhe.

Žižek, S. 2003. *On Belief.* London: Routledge. 52.

105

TECHNOLOGY AND THE END OF HISTORY FROM TIME CAPSULES TO TIME MACHINES
René Munnik

Time machines are not just imaginary devices that appear in science fiction. They exist. In fact, their precursors – time capsules – have been around for millennia. Since the last decades however, they turn into veritable machinery. The essential feature of a time machine is that it transforms non-contemporaries into contemporaries by eliminating their time distance. In order to understand how they work, let us have a brief look at the long history of time capsules. The most basic examples are simply relics of the past that convey fixated life signs from ages gone by – the kind of things palaeontologists and archaeologists search for in their endeavour to reconstruct lost worlds.

The invention of writing (c. 3000 BCE) and especially of the alphabet (c. 1000 BCE), improved time capsules considerably. They began to transfer articulated messages. This was the prerequisite for our common understanding of 'history' – a chronological complex of events witnessed by documents in archives and libraries. For centuries, history mainly comprised accounts of verbalised facts and thoughts, since the written word was its primary vehicle. Much later, in the nineteenth century, time capsules were improved once more: the introduction of photography, phonography, and film produced time capsules that contained fragmentary audible and visible traces of the past. They made it accessible for eyes and ears, and, by consequence, they changed our image of recent history fundamentally.

Today, in the age of information technology, we witness a revolutionary development of genuine time machines. Just like time capsules, they mirror our desire to foster the things we cherish or satisfy our curiosity – from scientific specimens to personal memories we treasure, up to the unesco's *World Heritage* and all the things that accompany the ethical and political imperative to commemorate. Apart from that, two things offer us the potential to actually realise them. The first one relates to our conceptions of knowledge and reality. The contemporary dominance of the information paradigm in our scientific understanding of the world means that 'real knowledge' of any object 'X' implies that 'X' is revealed as an instance or a product of some particular form of information processing. The second relates to our increasing technological abilities to realise anything describable as an instance or a product of information processing, by computers, gene-synthesis, and the like.

These two match perfectly. So, if you would know everything there is to know about 'X' – say, some human individual, its behaviour, its genetic makeup, its character, its intelligence, the sound of its voice, etc. – it would mean that you possess all the data and algorithms that characterise it. On the other hand, once you possess all those data and algorithms, theoretically you can duplicate it. The original and its reconstruction are more than perfect lookalikes. They are indiscernible, certainly from a scientific perspective, because all their relevant characteristics are identical. This is an instance of the boundary blurring of contemporary techno-science. It makes it gradually more difficult to draw the line between original and reconstruction, between reality and appearance, between nature and artefact, between man and machine.

Now, if you implement that into time capsules by making them transfer all the relevant data and algorithms, they become time machines. Time machines work by virtue of the blurring of a very particular boundary: the one between the (absent) things-gone-by and their (present) reconstructions. Consequently, the difference between the re-presentation of a past thing and its presence in the present (take these words in their temporal meanings) resolves. In old-time time capsules, you could easily distinguish the originals from their remaining traces. For instance the fossil of a T. Rex from the extinct species itself, or the image on a photo from the deceased person you miss. However, with state-of-the-art time machines you can have real T. Rex fooling around in some contemporary amusement park, or perhaps be disturbed by the uncanny presence of your late friend.

Only a very small part of it is realised these days and probably a lot of it will never be, but that does not mean that the whole idea is unrealistic. In fact, we are on the verge of collecting and documenting immense amounts of data that provide the necessary resources. Moreover, we are discovering more and more about the informational processes that determine reality, both biologically and psychologically. Finally, it is likely that we will possess the appropriate computational power to handle all these data. If so, we are becoming the inhabitants of a 'past' that will be pretty much of a present for those who live in the future.

In the remainder, I will draw on two issues. First, that the development of time machines is not a complete novelty, but rather the next stage in an age-old cultural trend. Secondly, I will point to its enormous societal and cultural impact, amounting to a complete transformation of our time constitution conceivable as 'the end of history'.

The Alphabet

Scholars in the humanities usually underestimate the role of the alphabet as the technological condition of literacy. Young children who are learning to read and write know this very well. However, once they have outgrown this preparatory phase, they enter a life world in which texts are omnipresent in religion, legal practice, science, art, education, and daily life – and so much so that literacy became an essential element of Christian and humanist anthropology. It all belongs to the enormous cultural spin-off of the technology of the alphabet.

Before the introduction of alphabetic writing, the word only existed as a spoken word (Greek: *mythos*). It was never without a mouth and a voice with a timbre, it was accompanied by gestures, and it demanded the proximity of speakers and listeners. In those days, people only knew about the more remote past from hearsay, from stories told: narratives, legends. It did not reveal itself by its own articulated and archived messages.

Stories that depended exclusively on memory and oral practices were susceptible to improvisation and contamination.[1] They were loosely fixated, and adapted themselves easily to changing circles of narrators and listeners. Even memorised and recited genealogies did, and if they changed, this could hardly be noticed because it lacked documents as benchmarks to discern it. By using metre, rhyme, and melody, the motility of words was restrained. These poetic means supported memory and served as a fixative for words. Stories, laws and prayers were preserved and perpetuated as songs, proverbs or poems. The name of the mother of the Muses – *Mnemosyne* – points at this didactic-societal role in oral traditions; it means '(collective) recollection.'[2]

Moreover, if a story were not kept in memory, by telling, performing, or singing it repeatedly, it would vanish without any trace. Only by appealing to their imagination and fascination, a story enticed people to retell it again and again. What perhaps began as a witness report of a battle would either transform into a heroic gigantomachy or might fall into complete oblivion. In a primary oral culture, the only account of the past consists of such once-upon-a-time type of tales. Its past is fundamentally different from history as we know it, the latter consisting of facts situated on a continuous timeline that runs into our own lifetime. It is 'mythical', and within an oral setting there is absolutely no alternative: it is *its* past without further qualification. Whoever, if any, is hidden behind 'Odysseus' remains shrouded in the mist of time. He certainly is not a historical figure, but his kind exemplifies the only mode of existence in which inhabitants of an orally transmitted past can appear. The decisive factor in the transition from non-history in prehistoric times to history as we know it, was the invention of the alphabet.[3] History *had* a beginning, and so, possibly it has an end.

Media

With some significant boosts – e.g. the rise of medieval universities, humanism in Renaissance, and the introduction of the printing press, the Reformation and the civilised urge for literacy in early Modernity –, the alphabet retained and strengthened its cultural monopoly until the 19th century. By then, other media were introduced, like photography in the 1830s. It was able to transfer much more information within the realm of visible things. The past began to expose itself

1 'Orality' here signifying 'primary orality', as coined by Walter Ong (2002), i.e. a total absence of literacy.
2 As Eric Havelock uses it in the title of his book The Muse Learns to Write (1986).
3 Brevity of exposition forces me to leave unmentioned other major consequences of scripture, like the rise of (literate) thought, religion (of the book), and (codified) law. For a somewhat more elaborate account, see Tijdmachines Munnik. 2013: 136-148.

as never before. Early photographers, well aware of the fact that photographic portraits counted as the successors of death masks, would advertise with slogans like 'secure your shadow ere the substance fade'. It enabled people to smile *post mortem* to the living with embalmed gestures. In general, the overall image of the past started to consist of journalistic pictures: photographic one-to-one representations.

When, some fifty years later, Thomas Alva Edison invented phonography, acoustic time capsules came around. While the alphabet unleashed words from ephemeral voices, Edison's machine unleashed voices from mortal mouths by capturing them on tin foil or wax cylinders; it tamed the goddess Echo. From then on, fragments of the past could resound on-demand, and singers could go on to affect the living, although their mouths and throats had long since perished in their graves. In the days before the invention of the phonograph, things were easy: if you heard a voice, you could be sure that someone was along. However, after that invention, it could also sound from a loudspeaker. Voices from loudspeakers can be equally from the living or from the dead, without you being able to hear any difference. Perishableness faded away from the realm of audible things. Consider this as a primitive forerunner of what time machines can do: that you experience something or someone, being unable to locate what you experience: something of the present, a trace from the past, something past, or all in once.

Without any doubt, past things will reveal themselves more and more sense appealing and lively in the near future. It already started in the 19th century and you might wonder: how vivid in the end? My answer: probably *very* vivid. Perhaps more than one would wish... an invasion of the inhabitants of *Hades*, who attained a *high-tech* high way into the present. Of course, you can demolish the whole machinery in order to silence them, but I doubt that you can justify such destructive act, because its elimination equals a book burning and can be considered committing high treason against history.

The End of the 'Historical Era'?

All time capsules are based on the same principle: something perishable is made persistent – spoken words, voices, statures, events are captured in writings, recordings, photos, and movies. Today, in neuroscience and Artificial Intelligence, the mind is mainly conceptualised in terms of information processing, and in molecular biology life is conceived in a similar way. They become recordable, storable, and reproducible as well.

Jurassic Park was an entertaining blockbuster. It was fiction, and probably it will remain so forever. Certainly, it contained some scientific blunders, but the basic idea was not just a fantasy. If you would possess all the genetic material of a past life form including the recipe to process it, you could revive it. Just like you could enjoy a performance of a lost Monteverdi madrigal if you found the original score in a forgotten archive. We do not possess enough genetic material of dinos to do it, but we have an abundance of such material at our disposal of still existing, endangered species. Currently, that is being stored massively in gene banks and frozen zoos, with the apparent purpose to avoid their demise. The environmental crisis and global warming gain apocalyptic proportions and catalyse an unprecedented urge for conservation: these gene banks have a status comparable with Noah's Ark. They are attempts to avoid at all costs what happened to the mammoth and the dodo: extinction. If it succeeds, the net result is that they will never become 'history' anymore.

These time machines are champions in a battle against the scandal of transience. They are attempts to undo the vanishing trick of time by chiselling the teeth from its jaws. The age-old development until now, suggests that it can succeed. The alphabet transformed the mythical time constitution into 'history' with its historical figures and facts. It is crucial for a 'historical fact' that it is *absent* and nonetheless counts as an *objective reality* because of its position on a timeline witnessed by writings that guarantee a continuous connection with the present. Oral cultures did not have that. They lacked the written witnesses and the continuity, and as a consequence the awareness of 'objective historical facts' did not even enter the mind. But literate minds definitely possessed such idea of objectivity, and used it as a standard to disqualify an orally transmitted remote past as 'mythical', henceforth meaning 'fictional'. Perhaps they are very profound fictions from a literary, psychological or existential perspective, but still fictions, anything but facts.

Time machines, in their turn, interfere in the historical time constitution that superseded the mythical version. They do so by substituting the continuous connection of writings by the formal identity of data and algorithms. By consequence the essential absence of 'historical facts' is superseded by re-presentations indiscernible from present realities. Figuratively speaking: 'a T. Rex in your backyard'.

We have become used to the fact that spatial distances do not really count anymore. A telephone brings you closer to someone abroad than to the man next-door. Distance is no longer about kilometres; it is a matter of the quality of mediating technologies. And the more sophisticated the medium, the more unnoticeable it becomes. In practice, it is irrelevant to object that what you hear in the phone is not your friend's voice far away, but its electromagnetic reconstruction from a headset. That is irrelevant because you are involved in a real conversation with your friend. Obviously, the world has shrunk.

Time capsules make the timeline shrink. People born after 1980 never knew John Lennon alive. Still, from their perspective, he did not pass away as remote as Enrico Caruso. Again, the relative temporal proximity of Lennon is not only a matter of the number of years since he died. It is realised through mediating technologies that happen to be available quite recently.

Time machines however, have the capacity to make the timeline implode altogether by teletransporting past things, no matter how far off temporally, to 'recency'. In doing so, they mess up the neat chronological sequence, and in doing so they blur the distinction between historians and journalists. And just like the telephone, it is irrelevant to object that they only make you experience an informational reconstruction, and not the past thing itself. That is because you are involved in a real confrontation with something that makes no difference with the past thing.

Our appreciation of past events, objects, people, anything, makes us destroy their pastness. We do not even have to prepare ourselves for this future post-historical human condition. It will just happen, though we can hardly imagine what it will be like. But if they live, people will get used to it, and when they do, perhaps they will be puzzled… not understanding how life was in a world in which things go by, just like we have difficulty in understanding how 'oralists' lived in a world exclusively conceivable as emerging from chimera's, hero's, demigods, and gods.[4]

109

4 Two remarks about '…go by'. The first: time machines only undo the transience of things of the object-kind, including appearances of other subjects. I do not make any claim concerning the undoing of mortality of subjects. Living under the post-historical human condition remains singular and may just as well be short. The second: entering a time machine implies isolating an item from its context. Consequently, particulars can be made persistent, but not their total context or 'world'. It is the strength of the hermeneutical approach that it emphasizes the historicity of the world rather than its components, as for example in Heidegger's Sein und Zeit 2006: 372-404: even in the post-historical condition, a lost world remains a lost world.

References

Havelock, E.A. 1986. *The Muse Learns to Write. Reflections on Orality and Literacy from Antiquity to the Present.* New Haven; London: Yale University Press.

Heidegger, M. 2006. *Sein und Zeit.* Tübingen: Niemeyer.

Munnik, M. 2013. *Tijdmachines. Over de technische onderwerping van vergankelijkheid en duur.* Zoetermeer: Klement.

Ong, W.J. 2002. *Orality and Literacy. The Technologizing of the Word.* New York: Routledge.

110

THE EMERGING POST-ANTHROPOCENE
Eric Parren

All Watched Over By Machines Of Loving Grace

I like to think (and
the sooner the better!)
of a cybernetic meadow
where mammals and computers
live together in mutually
programming harmony
like pure water
touching clear sky.

I like to think
 (right now please!)
of a cybernetic forest
filled with pines and electronics
where deer stroll peacefully
past computers
as if they were flowers
with spinning blossoms.

I like to think
 (it has to be!)
of a cybernetic ecology
where we are free of our labors
and joined back to nature,
returned to our mammal
brothers and sisters,
and all watched over
by machines of loving grace.

Richard Brautigan[1]

In 1967, the American novelist and poet Richard Brautigan wrote 'All Watched Over by Machines of Loving Grace', a visionary poem that paints a picture of a future world in which nature, human-kind, and cybernetic technology live together in harmonious coexistence. In Brautigan's vision, these machines of loving grace are computer systems that are fully integrated into our earth's natural systems, thereby creating a cybernetic ecologic techno-natural metasystem. This all-pro-viding system leaves humankind free to pursue its innate desires without the burden of labour and anxiety for survival.

When looking through these '60s San Fransisco-styled rose tinted glasses, the lineage to the present-day Silicon Valley induced rhetoric of trans-humanist Singularitarians is apparent. This techno-utopian vision has been passed down by the Whole Earth generation, and has slowly been absorbed into the broader cultural imagination, while along the way planting the seeds for its eventual realisation. A healthy dose of scepticism is indispensable in this regard, however, in recent years possible pathways towards the emergence of a global cybernetic ecology have become visible.

Technologies such as RFID, Bluetooth LE, and WiFi-Direct have enabled the develop-ment of devices that can be easily connected to larger planetary-wide cloud networks, ushering

111

1 Brautigan 1967.

in the era of the Internet of Things. By embedding sensors and computing technology in almost anything, from our home appliances, to live stock, and from globally shipped containers, to our cars and architecture, we are creating a networked smart-grid that collects data from innumerable sources and can actively respond to that data in real time. Noted futurist and sci-fi author Bruce Sterling sees this deployment of 'ubiquitous computing' as a one of the 'historic-scale ways to become cybernetically sustainable.'[2] Modern cities with their high concentration of networked devices, such as smart phones and IP cameras, are particularly well-equipped breeding grounds. Sterling observes that 'it's hard to imagine cities being denied a role in the sustainable and the cybernetic.'[3]

Media theorist Benjamin Bratton, borrowing a term from IT that describes a set of software and hardware that supplies the infrastructure for computing, has appropriately dubbed this planetary-scale system of computation we are fostering 'The Stack'. According to Bratton, we should not see 'the various species of contemporary computational technologies as so many different genres of machines, spinning out on their own', but 'instead see them as forming the body of an accidental megastructure.'[4] This megastructure is, as of yet, not evenly distributed, to paraphrase Gibson, but in our cities and transportation hubs its presence is palpable. Bratton defines six distinct layers in the Stack (Earth, Cloud, City, Address, Interface, and User) and proposes that 'the content of any one layer […] could be replaced (including the masochistic hysterical fiction of the individual User […]), while the rest of the layers remain a viable armature for global infrastructure.'[5] Arguably, we have already intentionally commenced on our trajectory to replace the 'User layer' quite some time ago.

Unknowable Intelligence
Since its inception in the 1950s, artificial intelligence has held great promise, and in the early days researchers in the field, such as Mavin Minsky, were confident that they would be able to create a true artificial intelligence within a few decades. However, after an initial spurt in the 1960s research activity died down, and apart from some very specific non-general applications for AI little progress was made during the following decades. Except for when the chess computer Deep Blue defeated world champion Garry Kasparov, in 1997, interest in the application and implication of artificial intelligence stayed mostly within the confines of computer science and sci-fi discourse. However, in recent years breakthroughs in specific fields of AI research have made headlines worldwide.

In 2011, when IBM's Watson was able to win a modified game of Jeopardy! against two former human winners, it signalled a big step forward in natural language processing, the process of extracting meaning from written words. Even though the DeepQA system that powers Watson may not be an actual thinking machine, it did demonstrate that its bio-inspired brain-emulating parallel computing framework is capable of processing massive amounts of data quickly and accurately.[6] What's more significant though, is that Watson showed that it was capable of learning on its own. Through its taxonomical and ontological processing, hypothesising, and evaluating abilities, it learned how to play Jeopardy! strategically. As journalist James Barrat describes in his analysis of Watson, after it 'got a correct answer in a category, it gained confidence (and played more boldly) because it realised it was interpreting the category correctly. It adapted to game play, or learned how to play better, while the game was in progress.'[7] The ability of a computing system to self-improve is a potentially disruptive notion given the increasing availability of computing power and the vast amount of data generated by sensor networks and human input on a daily basis.

Currently, the most striking examples of self-improving computer systems come from the field of machine learning, more specifically, the successes that have been booked with the advancements made in artificial neural networks. In 2014, Google acquired the start-up Deep-Mind that had created a remarkably successful self-learning system by combining artificial neural networks with reinforcement learning.[8] Both of these technologies are in essence cybernetic feedback systems that are inspired by biology. Artificial neural networks take inspiration from how our brain is structured and reinforcement learning mimics how organisms react to different stimuli. What is extraordinary about these approaches to machine learning is that the knowledge

2 Mulder 2010: 344.
3 Mulder 2010: 345.
4 Bratton 2014.
5 Ibidem.
6 Barrat 2013: 219-220.
7 Barrat 2013: 221.
8 Simonite 2014.

112-113 · Breeder is a software application that provides its users the ability to playfully explore the principle of artificial evolution. The software is based on the 'Biomorph' program as proposed by the British evolutionary biologist Richard Dawkins in his book *The Blind Watchmaker*. Variables like the colors, patterns, and movement of abstract visual elements are encoded into an artificial DNA. The user can crossbreed and mutate the genetic codes of these elements, thus creating new generations. This leads to an endless stream of rhythmically pulsating images that highlight the poetic beauty of the evolutionary process.

representation or 'intelligence' emerges from them without being explicitly implemented. Deep learning AIs are trained on large datasets of information, how they exactly make sense of this data inside of the neural network is often not clear, not even to the people who trained them. These are bottom-up approaches to giving a computer system the ability to self-improve by providing it with ample amounts of examples to learn from. Consequently, AI researchers have been baffled by some of the results that these systems have produced.[9]

DeepMind's first claim to fame came in December 2013 when it presented its software at a major machine learning conference. There, the company demonstrated how their system was able to play the Atari games Pong, Breakout, and Enduro better than any expert human player could. Leading AI researcher Stuart Russell recounted that 'people were a bit shocked because they didn't expect that we would be able to that at this stage of the technology […] I think it gave a lot of people pause.'[10] After being acquired by Google, DeepMind made international headlines in March 2016 by beating Lee Sedol, the world's top Go player of the last decade, in a five-game match, a task previously thought impossible for current-day AIs to achieve. The AlphaGo program that was used to play the Go matches was trained using DeepMind's deep learning technology and surprised its human opponent and supervisors with certain strategies it took that no human player would ever consider.[11] After the top-down approach of Deep Blue, and the parallel hypothesis generation and evaluation approach of DeepQA, the bottom-up up approach to implement AI championed by DeepMind, is quickly becoming one of the most successful method to develop new AI implementations.

While many of the noteworthy successes of AI have been when it was used to play games against human opponents, its real-world applications reach much further. Watson is currently used for medical diagnosing and deep learning algorithms are being developed to solve problems in numerous areas such as facial recognition, image search, real-time threat detection, analytics for finance and business purposes, voice recognition, motion detection, sentiment analysis, etc. A specific example is the current boom in research around driverless cars, which is in part propelled by the application of machine vision powered by neural networks. Both IBM and Google make computer chips based on newly developed architectures that are specifically designed to improve computing with neural networks.[12] [13]

113

This trend toward using self-learning and self-improving AIs sets us up for a dramatic set of circumstances. If we look at the bigger picture and see the application of artificial intelligence in relation to the rapid development of our planetary-scale computation stack, we can see how the data-producing systems we are creating on the one hand are increasingly being used in conjunction with the intelligence creating systems on the other hand. We are already giving our AIs access to the information produced by the sensor networks around our houses, in our cities, and on our farms. They are the Siris and Cortantas running on the servers connected to our mobile devices. They will be shuttling us around once driverless cars take over. Our drones fly on autopilot. Low latency algorithmic trading has fundamentally changed the way stock market works. Smart-grids powered by renewable energy sources are being implemented across the globe. The cloud layer of worldwide server networks that store all our data, all our knowledge, are being maintained with the aid of AIs. Clearly, we are deeply invested in this future.

The Post-Anthropocene

The implication of this is that we are starting to rely on systems of which we do not understand exactly how they do what they do and know what they know. Inventor and computer scientist Danny Hillis has labelled this new frontier that we are moving into 'The Age of Entanglement.' According to Hillis, 'Our technology has gotten so complex that we no longer can understand it or fully control it. […] Each expert knows a piece of the puzzle, but the big picture is too big to comprehend.'[14] Complexity scientist Samuel Arbesman states in agreement that we 'are in a new era, one in which we are building systems that can't be grasped in their totality or held in the mind

9 Ibidem.
10 Simonite 2014.
11 Silver and Hassabis 2016.

12 Jouppi 2016.
13 IBM 2016.
14 Arbesman 2016: 35-36.

of a single person; they are simply too complex.'[15] Arbesman anticipates that soon we will need to approach our own systems like we approach natural systems. We will 'need interpreters of what's going on in these systems, a bit like TV meteorologists' since 'we can't actually control the weather or understand it in all its nonlinear details', but 'we can predict it reasonably well, adapt to it, and even prepare for it'.[16] [17] In the near future humankind will once again be surrendered to a world over which we have little control. However, this time it will not only be nature that escapes our grasp, but we will live in 'a world where nearly self-contained technological ecosystems operate outside of human knowledge and understanding.'[18]

This sentiment is not necessarily new; versions of it have been around for over two decades. For instance, in 1997, scientific historian George B. Dyson already noted that 'we have mapped, tamed, and dismembered the physical wilderness of our earth. But, at the same time, we have created a digital wilderness whose evolution may embody a collective wisdom greater than our own. [...] We have traded one jungle for another.'[19] And, in 1999, N. Katherine Hayles in her discussion of the post-human declared that 'if the name of the game is processing information, it is only a matter of time until intelligent machines replace us as our evolutionary heirs.'[20] However, they were both arguing around the concepts of an artificial life form equipped with trans-human super intelligence that would succeed humans as the dominant intelligence on earth. In contrast, the planetary-scale computing system that we see emerging at the moment does not necessarily imply such a super intelligence, but rather points in the direction of a different kind of consciousness, a machine intelligence, or cybernetic ecology.

Even though we are just coming to terms with the fact that we have brought the earth into the Anthropocene, arguably, we might have to readjust that notion already within some decades considering that we as humans will not be the ones at the helm of spaceship earth anymore. The complex systems that we create and the algorithms that run them -the global cybernetic feedback networks of sensors, machines, and intelligences- are what is going to determine the future of our planet. The post-Anthropocene is emerging, it will be of our own making, and it may be here sooner then we expect. As Benjamin Bratton has observed: 'One of the integral accidents of the Stack may be an anthrocidal trauma that shifts us from a design career as the authors of the Anthropocene, to the role of supporting actors in the arrival of the post-Anthropocene.'[21] He also cogently mentions that 'The aporia of the post-Anthropocene is not answered by the provocation of its naming, and this is its strength over alternatives that identify too soon what exactly must be gained or lost by our passage off the ledge.'[22]

What the post-Anthropocene will look like is hard to predict, there are so many factors to consider, the complexity of the technology involved is increasing exponentially, and our discourse around the subject is only beginning to substantialise.[23] Hopefully, our collective efforts will not lead us towards a major catastrophe that will indirectly be the auto-termination of our species, but instead towards the utopian vision of a cybernetic ecology that Brautigan had, in which machines of loving grace do actually watch over us.[24]

15 Arbesman 2016: 19.
16 Arbesman 236.
17 Arbesman 2016: 179.
18 Arbesman 2016: 10.
19 Dyson 1997: 228.
20 Hayles 1999: 243.
21 Bratton 2014.
22 Bratton 2013.
23 An important technology that I've intentionally left out of this discussion is genetic engineering. The rapid advancements in technologies for gene sequencing, editing, and synthesising and the projected applicability of these technologies are indicative of the extreme significance genetic engineering will play in the future of the human race. The impact of genetically modified humans can hardly be understated and if the lofty goal of the end of ageing can be achieved the concept of what it means to be human will forever be changed. This in and of itself could be a form of post-Anthropocentic existence and thus would necessitate further investigation, but for the sake of conciseness I decided to keep this topic out of my analysis.
24 Barrat 2013: 229.

References

Arbesman, S. 2016. *Overcomplicated: Technology at the Limits of Comprehension.* New York: Current.

Barber, John F. 2016. Richard Brautigan: All Watched Over by Machines of Loving Grace. Accessed at: *brautigan.net/machines.html.*

Barrat, J. 2013. *Our Final Invention: Artificial Intelligence and the End of the Human Era.* New York: Thomas Dunne Books.

Bratton, B. H. 2013. Some Trace Effects of the Post-Anthropocene: On Accelerationist Geopolitical Aesthetics. e-flux journal. Accessed at: *www.e-flux.com/journal/some-trace-effects-of-the-post-anthropocene-on-accelerationist-geopolitical-aesthetics/.*

Bratton, B. H. 2014. The Black Stack. *e-flux journal.* Accessed at: *www.e-flux.com/journal/the-black-stack/.*

Dyson, G. B. 1997. *Darwin Among the Machines: the Evolution of Global Intelligence.* Cambridge: Perseus Books.

Mulder, A. 2010. The Future is Nothing but Ways Out, an interview with Bruce Sterling. In: *Politics of the Impure.* Rotterdam: V2_Publishing.

Hayles, N. K. 1999. *How We Became Post Human: Virtual Bodies in Cybernetics, Literature, and Informatics.* Chicago: University of Chicago Press.

IBM. 2016. IBM Research: Neurosynaptic chips. *IBM Research.* Accessed at: *www.research.ibm.com/cognitive-computing/neurosynaptic-chips.shtml.*

Jouppi, N. 2016. Google supercharges machine learning tasks with TPU custom chip. *Google Cloud Platform Blog.* Accessed at: *cloudplatform.googleblog.com/2016/05/Google-supercharges-machine-learning-tasks-with-custom-chip.html.*

Silver, D. & Hassabis, D. 2016. AlphaGo: Mastering the ancient game of Go with Machine Learning. Google Research Blog. Accessed at: *research.googleblog.com/2016/01/alphago-mastering-ancient-game-of-go.html.*

Simonite, T. 2014. Google's Intelligence Designer. *MIT Technology Review.* Accessed at: *www.technologyreview.com/s/532876/googles-intelligence-designer/.*

115

THE ETHICS OF LIFELOGGING 'THE ENTIRE HISTORY OF YOU'

Marleen Postma

> Wouldn't it be fantastic if a person could easily recall every moment of his or her life? Given that we live in such a technologically advanced society, why should a person ever have to forget what happened, when it was, where it happened, who was there, why it happened, and how he or she felt? If there was a system that could record everything a person sees, hears, and feels, how would that enhance our lives? We have enhanced our eye sight with prescription glasses, our hearing with hearing aids, and our timekeeping ability with watches. Enhancement of personal memories seems to be a natural next step.[1]

Even though the paragraph above reads as though it has been taken from a science fiction novel, it is in fact part of a text describing the University of Southern California's *Total Recall* project. The *Total Recall* project, which aims to design and develop such a personal information management system, is one of the most noteworthy research projects about lifelogging to date.[2] The definition of lifelogging that I will be employing in this paper is that formulated by Martin Dodge and Rob Kitchin:

> A life-log is conceived as a form of pervasive computing consisting of a unified, digital record of the *totality* of an individual's experiences, captured multi-modally through digital sensors and stored permanently as a personal multi-media archive. It will provide a record of the past that includes every action, every event, every conversation, every material expression of an individual's life; all events will be accessible at a future date because a life-log will be a searchable and recallable archive. Such a life-log will constitute a new, pervasive socio-*spatial* archive as inherent in its construction will be a locational record; it will detail everywhere an individual has been.[3]

The verb 'lifelogging' then refers to the act or process of capturing and storing the totality of one's experiences, thereby creating a personal, searchable archive of one's life.

After briefly assessing what benefits lifelogging could offer, I will investigate what the possible ethical risks associated with the practice of lifelogging are. Moreover, I will analyse in what way these risks are represented in the *Black Mirror* episode 'The Entire History of You' - an episode that revolves around lifelogging.

Risks and Benefits

There are multiple benefits associated with lifelogging. Throughout history, we have used various strategies and techniques to support our fallible, human memory. Lifelogging technology has the potential to provide each of us with a complete and searchable archive of everything we have ever experienced. However, lifelogs could be more than mere memory aids. One could argue, for example, that by preserving a record of all of our experiences, the lifelog preserves a record of us. In this sense, lifelogging constitutes a way to counteract the transience inherent in human life.[4] Similarly, O'Hara, Shadbolt and Tuffield propose that the lifelog constitutes a locus for the construction of identity. They go so far as to claim that 'the lifelog, for the lifelogger, might constitute the 'real' person'.[5] Whilst I would be cautious to equate a person's identity with their lifelog (which is a mere data set, rather than an embodied, living, feeling human being), I do think that the lifelog could be helpful in providing us with plenty of raw material from which to craft our life stories or narrative identities.[6]

116

1 Golubchik 2016.
2 Other noteworthy research projects into lifelogging are Microsoft Research's MyLifeBits project, which digitally chronicles the life of researcher emeritus Gordon Bell, and the LifeLog project of the American Defense and Advanced Research Projects Agency (DARPA).
3 Dodge and Kitchin 2005: 2. This definition is also used by Allen (2008).
4 Allen 2008: 54.
5 O'Hara, Shadbolt, and Tuffield 2008: 165.
6 Bell and Gemmell 2007: 60.

Moreover, lifelogs could help us evaluate our health and performance levels. Using predictive analytics, we could for example discover whether we are at risk of developing certain medical conditions and respond accordingly, e.g. through going to a doctor or by making lifestyle changes. It is here that we see a great deal of overlap with the Quantified Self movement, the slogan of which, 'self-knowledge through numbers', indicates its aim: for its users to keep track of and improve their health, fitness, and performance levels.[7]

Furthermore, lifelogging could increase accountability and personal security. As Anita Allen phrases it, 'a potential mugger or rapist would have to think twice in a society of lifeloggers'.[8] Having perfect records of everything and everyone would constitute a huge asset for law enforcement and security services. At the same time, it could help protect innocent civilians from security forces by recording any abuses of power.

Many of these benefits would, however, require the sharing of our personal lifelog data with others, whether it be health care professionals, security services, or the companies providing us with lifelogging technologies. It is here that the majority of privacy concerns arise.[9] Most of the ethical risks involved in lifelogging can be subsumed under two categories introduced by Anita Allen: pernicious memory and pernicious surveillance.[10] The notion of pernicious memory refers to the downsides of having a perfect memory machine. It is important to point out, as Allen does, that encountering the record of a past experience need not cause one to literally remember that experience; the record can merely serve as a prompt.[11] It is more important, however, to stress that a perfect memory is not an unqualified good. In fact, attention is increasingly being paid to the importance of forgetting as a necessary complement to memory.[12] Having a detailed archive of the totality of our experiences could enable excessive rumination as well as the dredging up of the past.[13] In order for us to move on, it may be crucial for both others and ourselves to forget the past, or at least to lay it to rest.

The benefits of increased accountability and personal security would most likely come at the cost of increased surveillance. As both Dodge and Kitchin and Allen point out, lifelogging has both sousveillance and surveillance dimensions. It is sousveillance in the sense that our lifelog would capture data about ourselves or from the perspective of ourselves. It is surveillance, in the sense that our lifelogs would also capture data about others, including others who may also be engaged in acts of sousveillance or surveillance.[14] Lifelog data will need to be accessible and useable in order for us to reap its benefits, thus creating the risk of other people, either legally or illegally, accessing our private data.

However, the most significant ethical risk of lifelogging, I would argue, is its potential to change who we are, both on an individual and on a societal level. The sheer knowledge that someone could access our private data can influence our behaviour.[15] If lifelogging were to become a common practice, this could influence our reasonable expectations of privacy, as well as the potential for trust between people. As it is challenging for us to imagine what issues might arise in a society of lifeloggers, I will now turn to an artistic representation of such a society: the *Black Mirror* episode 'The Entire History of You'.

'The Entire History of You'

'The Entire History of You', which revolves around protagonist Liam Foxwell, a young lawyer, starts with Liam's work appraisal meeting. After leaving the meeting and getting into his car, Liam replays his recording of the meeting from his Grain, his lifelogging implant, and dwells on the parts of the appraisal that did not go so well. He subsequently arrives at an airport, where he is asked by security to fast forward through his recordings from the last week. We then see Liam arriving at a dinner party hosted by his wife Ffion's friends, where he witnesses his wife talking to Jonas, a man he has never met before. Ffion's friends enquire about Liam's appraisal and suggest

7 Quantified Self 2016.
8 Allen 2008: 52.
9 As Anita Allen points out, the act of capturing data itself also implicates privacy concerns, especially if the recording contains expressions of the lives of other persons. Allen 2008: 54-55. Moreover, lifelog data could potentially be illegally obtained, for example by hacking.
10 Allen 2008: 55.
11 Allen 2008: 56.
12 See in this regard Bannon, 'Forgetting as a Feature, not a Bug',

Dodge and Kitchin, 'The Ethics of Forgetting in an Age of Pervasive Computing', Mayer-Schönberger, Delete: The Virtue of Forgetting in the Digital Age and in this volume: Van Bree, 'Digital Hyperhymesia: On the Consequences of Living with Perfect Memory'.
13 Allen 2008: 56-65.
14 Allen 2008: 54.
15 One can witness this phenomenon on social media already: once people realise others (for example possible employers) might be monitoring their online behaviour, they often adjust their behaviour and self-censor.

replaying the meeting for them all to see, so they can 'appraise the appraisal'.[16] However, Jonas steps in to save Liam from the embarrassment and the group sits down for dinner. During dinner, Jonas speaks very frankly about his personal life; he mentions that he sometimes masturbates to 're-dos' of 'hot times' from earlier relationships.[17] Liam becomes increasingly suspicious of the way his wife looks at Jonas and laughs at his jokes. Together, the group of friends watches some Grain recordings from several years ago.

Back at home, Liam and Ffion watch the recordings from their baby daughter's Grain, taken when the baby was at home with the babysitter. During their conversation, it turns out that Jonas is in fact 'Mr Marrakesh', someone Ffion dated in the past.[18] Ffion had previously down-played her relationship with 'Mr Marrakesh', claiming it had only lasted a week, but from the Grain footage that was played at the party, Liam discovered that Ffion and Jonas in fact dated for much longer. The couple gets into a row, but they ultimately make up and have sex (while both watch-ing 're-dos' of more passionate moments from earlier in their relationship).

After they finish, Liam returns to the living room to watch more of his Grain's recordings of the dinner party, all the while drinking heavily. He continues doing this all night and also uses lip-read reconstruction software to find out what Ffion and Jonas said to each other at the party. Liam gets increasingly worked up and becomes convinced that something has been going on between Jonas and his wife. Ignoring his Grain's warnings that he is not fit to drive, Liam gets in his car and drives over to Jonas' house to confront him. Under threat of violence, Liam then forces Jonas to show his Grain's recordings of Ffion on a TV screen and to wipe them all.

Next, Liam wakes up having driven his car into a tree. He walks home, where he con-fronts Ffion with something he found out from Jonas's Grain: the fact that Jonas and Ffion slept with each other some 18 months ago, around the time their daughter was conceived. Liam forces Ffion to show her recordings of her and Jonas, in order for her to prove that Jonas wore a con-dom and Liam is indeed the father of her baby. The closing scenes show Liam walking around his now empty house, replaying recordings of happier times with his wife and daughter, before he goes into the bathroom to cut out his Grain with a razor and cuticle clippers.

The problems of pernicious memory and pernicious surveillance are clearly depicted in this episode. We witness Liam's excessive rumination over his appraisal meeting. Moreover, Liam's Grain enables him to 're-do' moments from the dinner party over and over again, fuelling his suspicions and mistrust of his wife. Even though it turns out that Liam's suspicions were justified, he is none the happier for it. The risks of dredging up the past, too, are strongly represented. In his job appraisal, Liam asks whether 'we', i.e. the law firm, are 'morally okay' with litigation in retrospective parenting cases: cases where children sue their parents for damages caused by mistakes or shortcomings in their upbringing.[19] From his tone and facial expression, it is clear that Liam has his doubts about the practice. The scene in which Liam replays footage of himself threatening Jonas with violence, too, invokes the risk of dredging up the past, as the viewer knows that this event, which was also record-ed by Jonas' Grain, could seriously damage Liam's career and even his entire future.

This last example brings me to the problems of pernicious surveillance and threats to privacy. Liam's lifelog data is not only accessible to him; it will also be accessed by his employer if his contract is to be renewed, as well as by airport security whenever he needs to take a flight. Furthermore, we witness Ffion's friends softly pressuring Liam to 're-do' his interview with them. As soon as Liam and Ffion come home from the dinner party, they replay the recordings from their daughter's Grain to make sure nothing strange happened while they were away. Their baby daughter is completely unable to deny them access to her Grain.

Most importantly, however, we see that lifelogging can change people and the rela-tionships between people. In a world where everyone has a lifelog and where these lifelogs are accessible by others, both our individual privacy and the privacy of our relationships is compro-mised. The fact that everything we (and everyone around us) have ever done is on record can seriously damage our expectations of privacy, as well as the relations of trust between people. As he is reviewing footage from the dinner party, we see Liam becoming more and more paranoid.

16 'The Entire History of You', Black Mirror, season 1, episode 3.
17 Ibidem.
18 Ibidem.
19 Ibidem.

At one point, Liam exclaims: 'This isn't me! Look at what you're doing to me!' - where 'you' refers to Ffion.[20] The viewer knows that rather than Ffion doing this to him, this is what Liam does to himself under the influence of his Grain.

Reflection

'The Entire History of You' clearly represents the risks of pernicious memory and pernicious surveillance. Furthermore, it shows us what could happen to people and to interpersonal relationships in a society of lifeloggers. Of course, 'The Entire History of You' is an artistic representation of a society in which lifelogging is a widespread, accepted phenomenon; it is not a real world scenario. I do, however, think that such a representation can encourage viewers to think about the ethical risks of lifelogging and possibly help them to better understand these risks. In a sense, 'The Entire History of You' thus serves as a warning to its viewers.

As I mentioned before, ever more scientists are acknowledging the importance of forgetting in the digital age. Bannon, Dodge and Kitchen and van Bree are exploring ways in which forgetting could be built into lifelogging technology. Allen, on the other hand, focuses more on the laws, rights and regulations that could support the safe implementation of lifelogging into society. Lifelogging in the strict definition that is employed in this paper may never come into being on a large scale. Nevertheless, some aspects of lifelogging are already implemented in today's society – think for example of how our mobile phones already generate a locational record, or of all the digitised memories we have posted on social media – and more are likely to follow. 'The Entire History of You' shows us how pervasive lifelogging technology could be and urges us to reflect on the possible impact of lifelogging technology on the way we live our lives. As everyone would be affected in a society of lifeloggers, it is crucial that not only those developing the technology, but also lawyers, philosophers, artists, and the general public reflect on how we can best benefit from lifelogging technology without falling victim to its risks.

119

References

Allen, A. 2008. *Dredging up the Past: Lifelogging, Memory and Surveillance.* University of Chicago Law Review, 75: 1, 47-74.

Bell, G. and Gemmell, J. 2007. *A Digital Life.* Scientific American, 296: 3, 58-65.

Dodge, M. and Kitchin, R. 2005. *The Ethics of Forgetting in an Age of Pervasive Computing.* CASA Working Papers Series, 92, 1-24.

Golubchik, L. 2002. *Total Recall: a Personal Information Management System.* Internet Multimedia Lab at the University of Southern California, *bourbon.usc.edu/iml/recall/* website accessed 9 August 2016.

Mayer-Schönberger, V. 2009. *Delete: The Virtue of Forgetting in the Digital Age.* Princeton University Press.

O'Hara, K., Shadbolt, N. and Tuffield, M. 2008. *Lifelogging: Privacy and Empowerment with Memories for Life.* Identity in the Information Society, 1:1, 155-172.

'The Entire History of You.' Black Mirror, season 1, episode 3, Endemol UK, 18 December 2011. Netflix, available at: *www.netflix.com/watch/70264856?trackId=13752289&tctx =0%2C0-%2Ce23663bc-b998-4913-9170-1feb1c097d6a-6631378.*

Van Bree, T. 2016. 'Digital Hyperthymesia: On the Consequences of Living with Perfect Memory.' *The Art of Ethics in the Information Society – Mind You,* Janssens, L. (Ed.), Amsterdam University Press, 28-33.

Quantified Self. Quantified Self Labs, 2015, quantifiedself.com. Website accessed 9 August 2016.

PRIVACY AS A TACTIC OF NORM EVASION, OR WHY THE QUESTION AS TO THE VALUE OF PRIVACY IS FRUITLESS

Bart van der Sloot

Privacy aims at avoiding norms, whether they be legal, societal, religious or personal. Privacy should not be regarded as a value in itself, but as a tactic of questioning, limiting and curtailing the absoluteness of values and norms. If this concept of privacy is accepted, it becomes clear why the meaning and value of privacy differs from person to person, culture to culture and epoch to epoch. In truth, it is the norms that vary; the desire for privacy is only as wide or small as the values imposed. It can also help to shed light on on-going privacy discussions. The 'nothing to hide' argument may be taken as an example. If you have nothing to hide, so the argument goes, why be afraid of control and surveillance? The reaction has often been to either argue that everyone has something to hide, or to stress why it is important for people to have privacy. For example, it has been pointed out that people need individual privacy in order to flourish, to develop as an autonomous person or to allow for unfettered experimentation. This; however, is, in general, a rather weak argument. How, for example, has the mass surveillance activities by the NSA undermined the personal autonomy of an ordinary American or European citizen? Moreover, many feel that national security and the protection of life and limbs is simply more important than being able to experiment unfettered in private. The rhetorical question "Who needs privacy when you are dead?" is often asked. This essay will argue that there may be a stronger argument to make when the focus is turned around, namely not by looking at privacy as an independent value, which might outweigh or counter other interests, but as a doctrine which essence it is to limit and curtail the reach and weight of other values.

To illustrate this point, this contribution will focus primarily, though not exclusively, on the evasion of public and legal norms, but the arguments made are no less valid for other types of rules.[1] This short contribution will give three examples of theories focusing on the freedom from the application of the public norm. Far from being exhaustive, three archetypical strands of this tactic will be discussed: the biblical Garden of Eden, the Arendtian interpretation of the Aristotelian household and the Lockean compromise between the tyranny of the patriarch and that of the monarch. In the first, there is no need for privacy as there is no morality. In the second, norms and laws exist, but cannot be applied to the private domain. In the third, norms and laws could be applied to the private sphere, but there is no need to. What they have in common is that they can all be viewed as tactics of limiting or curtailing the reach of the public norm.

The Biblical Garden of Eden

An example of the dialectic relationship between norm and privacy can be found in the biblical Garden of Eden. When God made man and woman, 'they were both naked, the man and his wife, and were not ashamed.' The absence of shame signals an absence of both morality and privacy. Shame is the particular feeling man begets when not living up to a norm, either imposed by himself, by society, by the state, or by God. It is also the reason why animals, for whom there is no difference between is and ought, have neither shame nor a desire for privacy. Shame is always the mediating factor between norm and privacy. It is only in the next chapter of the bible that a rule or norm is introduced, namely the command of God not to eat from the tree of Good and Evil. It needs not be recalled that at this point, only the Lord has moral understanding, which makes him the ideal lawmaker but his subjects become transgressors by necessity. Without knowledge of good and bad, man is not only unaware of law's underlying morality, but also incapable of grasping that violating God's command is wrong.

---

1 In general, three tactics of norm-evasion can be distinguished, namely by claiming a form of (1) negative freedom, (2) autonomy or (3) positive freedom. In the first category, privacy is deployed as a tactic of norm evasion by installing zones of non-normativity. In theories that describe privacy as enhancing personal autonomy, privacy functions as a tactic of norm evasion by creating zones of private normativity. Here a formal power is let or transferred to an individual or group to engage in private norm setting. Finally, in theories that describe privacy as a right to pursue a certain positive freedom, the public norm is opposed by a value-laden concept.

When Adam and Eve eat from the tree and beget knowledge of good and bad, imme-diately a sense of shame and a desire for privacy arises: 'the eyes of them both were opened, and they knew that they were naked; and they sewed fig leaves together, and made themselves aprons.'[2] In the biblical story, it is not the violation of the norm Adam and Eve are afraid of, as they are themselves the trespassers, but the fact that they understand that they have violat-ed the norm. Particularly, it is their own nakedness, their own corporality, which is shameful to them; they cover their body and the Lord later clothes them with animal skins to hide their animal descent, of which their body is a remnant. God bans Adam and Eve from paradise out of fear that they would also eat from the tree of life. Man is now forever doomed to dangle between beast and God. Like God, man possesses knowledge of good and bad, while, like animals, remaining mortal. Like animals, man has bodily needs in order to keep him alive, both as an individual (eating, sleeping, defecation) and as a species (procreation), and similar to God, man has intellectual capacities enabling him to engage in rational discussion, have a moral un-derstanding, and create norms and laws. In the history of mankind, this dual nature has always led to an inevitable tension, as man, either through *Imitatio Christi* or some other doctrine, has been urged to sublimate his first nature and substitute it for or suppress it by his second. The first nature is hence always a source of shame and is hidden in a private domain; it is their body, their animal descent, that shames Adam and Eve and that they try to hide. The body is and remains a pre-moral place.

The Arendtian Interpretation of the Aristotelian Household
This duality in nature was also at the centre of political thought in Greek philosophy. Similar to other creates (*zoé*), man was thought to have natural desires and needs, but as a *zoōn politikon*, man possessed *logos* and was inclined to live in political communities, ruled by reason and jus-tice. According to Aristotelian ethics, to reach a state of happiness (*eudemonia*), man needed to reach a state of self-sufficiency (*autarkeia*), satisfy his intellectual needs, and long for complete-ness (*teleiotēs*). For both Plato and Aristotle, the *polis* and its laws were necessary to fulfil man's second nature, hence Plato's rejection of the city of pigs and Aristotle's thesis that the city-state comes into being for the sake of life but exists for the sake of the good life. Both also seemed to think that for exceptional men, philosopher kings, the god amongst men or the *megalop-suchus*, it was possible to fully sublimate their first nature. Where both differed, however, was on the question of whether amongst less godly men, the first nature could be suppressed or sublimated though the guidance of the *polis* and its laws.

Plato famously thought it could, at least in the warrior class, for which all life was com-munal. The philosopher-kings were essentially destined to concern themselves with the rational side of man, adopt laws, and impose them on the rest of the community. The working class was engaged in providing the community with the necessities of life, such as food, shelter, and basic products. The warrior class was installed to preserve the community, both by protecting it from physical danger and trough procreation. This communal essence led Plato to propose a life in which all meals, spouses, children, and property were shared, subjected to the communal interest formulated by the rules of the philosopher kings. The Aristotelian rejection was based on two communicating arguments. First, Aristotle renounced the Platonian suggestion of one communal interest, to be determined by the philosopher king. Rather than a unity, he believed the public domain to be a plurality of opinions of equal value. Second, the private domain was essentially a sphere in which the private interest prevailed. Although in a sense, all men have the same interest in survival, the shared interest in not a common interest but a private one. The private interest in self-sufficiency naturally prevails over the common interest: people take better care of their own interests, their own property, and their own children.

On this Aristotelian principle, the Arendtian concept of private and privacy is based. Arendt alludes to the sharp distinction between what was private (*idiom*) and what was com-munal (*koinon*) and to the separation of the household, as a place where self-sufficiency and *au-tarkeia* could be reached, from the public domain, where man's second nature could be fulfilled.

122

2 Genesis 3:7.

For Arendt, this distinction is essentially one between necessity and freedom. Man's first nature imposes on him certain needs, which he cannot choose not to fulfil.

> The distinctive trait of the household sphere was that in it men lived together because they were driven by their wants and needs. The driving force was life itself which, for its individual maintenance and its survival as the life of the species needs the company of others. That individual maintenance should be the task of the man and species survival the task of the woman was obvious, and both of these natural functions, the labor of man to provide nourishment and the labor of the woman in giving birth, were subject to the same urgency of life. Natural community in the household therefore was born of necessity, and necessity ruled over all activities performed in it.[3]

In contrast, freedom and justice dominate public life. Man, relieved from the necessities of life, enters the domain of plurality and choice. Normativity dominates the public domain because it is here that man's second nature — his rational and moral capacities — flourishes. The other way around, as only this domain is one of freedom and plurality, it is only here that man could do good or bad and that laws and norms are needed to guide behaviour. It is only when man has a choice that laws and norms have relevance. The household, being a place of necessity, is not subjected to these laws; thus in contrast to the Garden of Eden, norms do exist, but they *cannot* be imposed on the household. According to Arendt, what the Greek philosophers 'took for granted is that freedom is exclusively located in the political realm, that necessity is primarily a prepolitical phenomenon, characteristic of the private household organization, and that force and violence are justified in this sphere because they are the only means to master necessity — for instance, by ruling over slaves — and to become free.'[4]

The Lockean Compromise Between the Tyranny of the Patriarch and That of the Monarch
Locke's first of his Two Treatises on Government denounced Filmer's Patriarcha, in which he argued that God had transferred to Adam, e.g. the male, the natural and absolute authority to rule over the family, that this family was in fact a small commonwealth and that the only exception to the rule of the monarch over the greater commonwealth is that the patriarch over his family, which included the power to decide over life and death of his subjects. Locke's second treatise gives a description of the state of nature and the contract installing the political state, which, as far as relevant here, offers an alternative to Hobbes's Leviathan. For Hobbes, the natural condition was not one of authority and hierarchy, such as with Filmer, but one of radical equality, as every man and woman had but one concern: self-preservation. Famously leading to a war of all against all, the monarch was to be granted an absolute power. In contrast to Filmer, Hobbes let no private authority to the patriarch. The only room left to individual autonomy was where the public laws where silent (*silentium legis*), which was not an individual right but subject to the discretion of the monarch.

123

Locke denounced both positions, wanting to retain a private sphere but without the looming inequality and injustice following from Filmer's position. He did so, essentially, by treating the household as a peaceful society, where public norms and political authority could, but need not be applied.

> Demonstrating that relations in the household can be peaceful, even if the power of life and death is not assigned to the adult male governing it, solves two problems at once. It proves that there exists at least one institution, the family, in which adults need not be controlled by an all-powerful rule; and it addresses the question of conflicting jurisdictions. If the power which the heads of households need to ensure order is not political, it will be compatible with the power of the magistrate, and heads of households will be entitled to maintain it even within the political association.[5]

3 Arendt 1998: 30.
4 Arendt 1998: 31.

Locke's argument goes as follows. Having first denounced any form of natural authority such as proposed by Filmer, Locke's starting position was very much egalitarian, like that of Hobbes. To change a Hobbesian state of nature, he introduced, besides the duty of self-preservation, the duty of preserving the human race. Man's first and natural society, the family, is especially concerned with this second duty, as it is installed, according to Locke, not merely with an eye on procreation, but the continuation of the species as a whole. Although this first society creates a paternal power, this does not come into conflict with the later installed political power. As 'these two Powers, Political and Paternal, are so perfectly distinct and separate; are built upon so different Foundations, and given to so different Ends',[6] the paternal power is retained even in a political society. While the end of political society is the preservation of man, through the protection of life, liberty, and estate, the end of conjugal society is the preservation of mankind. While the first interest, self-preservation, as with Hobbes, leads to conflict and necessitates a civil society and political power, the second interest does not.

The essential difference between the conjugal and the political society is that in the first, political power is absent, which is defined by Locke as the power of law enforcement through punishment, essentially the power over life and death. The parents have 'no Power of Governing, i.e. making Laws and enacting Penalties on his Children',[7] nor does the man have such power over his wife. Not only is private normativity and subsequent norm enforcement absent in the family, according to Locke, it is also a place of equality. Man and woman enter the conjugal society voluntarily and marriage is subjected in total to the contractual terms upon which they agree. Likewise, although not fully equal in factual terms, the child is under no obligation to obey his parents. His subjection to the parental authority is based on his (tacit) consent. The family remains this peaceful and harmonious place because of two reasons.

First, Locke proposes that if the private domain does become one of conflict, the family is simply dissolved. As man and woman have entered into marriage voluntarily, they have agreed on the terms and conditions. If these terms of are violated, either one is free to dissolve the marriage. The same applies to any children. Although there might be an authority vested in the parents by the child, this gives them power only in so far as this is to the child's benefit. If they violate this rule, the terms of the tacit contract will be violated and the child is free to withdraw his consent. Second, and more important, Locke's description of the household is essentially one of paradisiacal peace. Although all family members are equal, they are not identical; rather, all are dependent upon each other. The male member of the family is dependent on his wife to give birth to their child and the woman is dependent on man because what separates humans from other animals, which also explains why humans have such long-lasting marital relations, is the fact that 'the Female is capable of conceiving, and *de facto* is commonly with Child again and Brings forth too a new Birth long before the former is out of a dependency for support of his Parents help.'[8] Hence the father is needed to take care of both women and child. The child, in turn, is dependent on its parents for support, and vice versa, the parents wanting to fulfil their second natural duty must provide the child with protection and education. All members of the family thus understand that they are dependent upon each other and normally, conflict and coercion will be absent as a consequence. Consequently, public norms and political authority could hypothetically be applied on the household, but there is no reason to.

On the Value of Privacy

This contribution has suggested that one of the reasons why there is no dominant theory or explanation on the value of privacy and why the value of privacy differs from person to person, culture to culture, and epoch to epoch is that privacy has no value. Its essence is precisely that of limiting and curtailing norms and values; privacy is a tactic of norm evasion. The suggestion is that there is no unifying or common element in the various privacy theories that have been put forward so far, other than that they represent the other side of the coin of norms and values. The commonality between these theories is precisely that privacy is directed at curtailing and limiting norms. This might explain why the meaning and value of privacy seems to be a cha-

124

5 Gobetti 1992: 67.
6 Locke 1998: §71.
7 Locke 1998: §74.
8 Locke 1998: §80.

meleon; it is not privacy that keeps on transforming, it is the norms that do so. It would take a book to exhaustively demonstrate this hypothesis, this contribution has merely made a small start by describing three theories in which privacy, the body and the private domain, serve as a place of non-normativity.

References

Arendt, H. 1998. *The Human Condition.* Chicago: The University of Chicago Press.
     30-31.

Gobetti, D. 1992. *Private and public: Individuals, households, and body politic in Lock and Hutcheson.* London and New York: Routledge. 67.

Locke, J. 1998. *Two treatises of Government.* Cambridge: Cambridge University Press.

HOW ART CAN CONTRIBUTE TO ETHICAL REFLECTION ON RISKY TECHNOLOGIES
Sabine Roeser

Debates about technological risks are highly complicated as they involve scientific and moral uncertainties and complexities. Where technologies are often developed to improve people's wellbeing, society is frequently not convinced of the necessity and desirability of such developments. Current approaches of conducting debates about technological risks are often not fruitful. They are frequently heated and culminate in stalemates. This indicates a deeply rooted mistrust of risky technologies and a problem of legitimacy in democratic society. Other approaches are needed to conduct debates, for example involving visual arts and literature.

In this essay I will examine the contribution that art can make to improve public deliberation concerning technological risks by engaging people in emotional-moral reflection. This is an unexplored field that can make a major contribution to academic research as much as to public debates about risky technologies. The central idea to be investigated in this essay is that art that reflects on and engages with risky technologies, i.e. 'techno-art', can make abstract problems more concrete, aid in the exploration of new scenarios, challenge the imagination, and broaden narrow personal perspectives through empathy, sympathy, and compassion. This way, techno-art can enhance the quality of public deliberation and decision making with regard to technological risks, help to overcome stalemates and lead to morally better technological innovations.

Techno-Art
New technologies are often severely debated in society. Due to environmental concerns and experiences with accidents, many people have a critical stance towards new technologies, which are experienced as risky. On the other hand, artists have become more and more interested in these technologies. There are numerous artists who are exploring the possibilities as well as the controversial aspects of new technologies in their work. This is what I call 'techno-art': art that reflects on and engages with risky technologies.

Precursors of Contemporary Techno-Art
There are historical examples of artists who have engaged with scientific and technological developments, most famously Leonardo da Vinci. However, many artists and writers did not even touch upon the subjects of science and technology, whereas others took a critical stance. Romanticism for example idealised a pre-industrialised world. Dystopian novels examined the possible risks of technological developments, and they still speak to the imagination. For example, Aldous Huxley's *Brave New World* and Mary Shelley's *Frankenstein* articulate and have shaped people's suspicions towards, for example, cloning and human enhancement. On the other hand, there have also been more utopian artistic movements that did explicitly embrace technology. For example, several modernist movements at the beginning of the twentieth century celebrated technology and the efficiency and rationality it encompasses. For example De Stijl, Constructivism, and Futurism have been inspired by these developments. Science fiction is a well-established genre in literature and film dealing with utopian as well as dystopian visions of technological developments and their possible impacts on society, probing ethical reflection with aesthetic means and via the imaginative and emotional capacities of the audience. Science fiction author Isaac Asimov developed ethical guidelines or so-called laws of robotics implicitly and explicitly in his science fiction novels.[1]

Contemporary Techno-Art
There are currently quite a number of artists and writers engaging with new technologies. There are new literary genres devoted to environmental as well as to climate issues, and there are even specialised academic journals devoted to the study of these new genres. Prominent writers Kazuo

127

1 Cf. Asimov 1950.

Ishiguro and Michel Houellebecq have written novels that explore a world in which cloning is a commonly established practice.

An increasing number of visual artists also engage with new technologies. These new technologies provide for new material possibilities as well as opening a whole new array of topics to engage with. Examples are nuclear artist William Verstraeten, climate artists David Buckland and Boo Chapple, and artists engaging with ICT and robotics, e.g. Stelarc, bio-artists Eduardo Kac, Jalila Essaidi, and Patricia Piccinini.

Working on the intersection of bio-art, robotics-art and ICT-art, in 2007, performance artist Stelarc experimented with his own body by attaching a third ear to his arm by surgery and cell-cultivation, partially using his own stem cells. Eventually, there will be a microphone and a wireless connection to the Internet implanted into the ear. On his website, Stelarc explains the idea as follows:

> For example, someone in Venice could listen to what my ear is hearing in Melbourne. [...] Another alternate functionality [...] is the idea of the ear as part of an extended and distributed Bluetooth system. [...] This additional and enabled EAR ON ARM effectively becomes an Internet organ for the body.[2]

With this work, Stelarc explores the possibilities of stem cell research and enhancement in a way that goes beyond the ways in which contemporary scientists usually approach such developments. He does it in an imaginative, playful, and provocative way, exploring the technological and scientific possibilities and their legal and ethical boundaries. It is not directly clear what the use of EAR ON ARM would be, which is why scientists and technology developers would probably not even think about something like this. Rather, it explores conceptual and normative issues, for example the meaning of connectedness in an age of hyperconnectivity.

'The Young Family' by Patricia Piccinini is a sculpture of a human-animal hybrid. It shows a mother and her infants in their bare flesh who look part human, part pig, part dog, and are described by one critic as 'monstrous cute.'[3] While looking strange and repulsive, Donna Harraway writes about the care that Piccinini's creatures evoke. She sees them as symbolic for our modern technoculture in which we are obliged to care for what has been artificially created:

128

> Natural or not, good or not, safe or not, the critters of technoculture make a body-and-soul changing claim on their 'creators' that is rooted in the generational obligation of and capacity for responsive attentiveness. [...] To care is wet, emotional, messy, and demanding of the best thinking one has ever done. That is one reason we need speculative fabulation.[4]

Harraway here refers to the emotions that are invoked by the artwork, but she also emphasises the need for critical thinking emanating from this emotional experience.

Techno-Art and Emotional-Moral Reflection About Risk
The examples discussed above indicate that techno-artworks can reach out to people in a very direct way, by drawing on their imaginative and emotional capacities. Works of techno-art provide for concrete, vivid images and narratives that make technological developments and their possible impact on society more tangible for a larger audience than is possible via highly specialised scientific articles and abstract philosophical theorising.[5] Techno-artists can create awareness of societal issues,[6] and provide for critical reflection on technological and scientific developments.[7] This is an area that is still largely unexplored by philosophers and scholars who study societal aspects of risky technologies.

Referring to artworks more in general, so not focusing specifically on techno-art, philosophers have developed accounts on how art helps our ethical and political thinking.[8] These philosophical approaches have mainly focused on more traditional works of art that engage with interpersonal and societal relationships. However, works of techno-art are materially and content-wise different from such works of art, giving rise to different philosophical questions, for example: which aspects of

2 See stelarc.org/?catID=20242.
3 Goriss-Hunter 2004.
4 Harraway 2007.
5 Cf. Zwijnenberg 2009: 19.
6 Gessert 2003.
7 Reichle 2009, Zwijnenberg 2009.
8 E.g. Caroll 2001, Nussbaum 2001, Gaut 2007, Adorno et al. 1980, Rorty 1989, Kompridis 2014.

technological risk are addressed in the artwork? Which moral values are involved in the technology exemplified or used in the artwork? How does the artwork address or highlight these values?

Furthermore, an important aspect that is not yet discussed in the (limited) academic literature on techno-art is that such art-works often give rise to emotional responses. This gives rise to the question as to which role emotions might play in ethical reflection triggered by works of techno-art. For example, the work of Stelarc might give rise to fascination about the possibilities of technology, as well as to disgust at the gross image, or annoyance at the seemingly decadent endeavour. What kind of values and concerns do these emotions point to, and how should we assess them? The work by Piccinini gives rise to feelings of care as well as to feelings of discomfort and uncanniness. The latter can point to the unclear moral status of artificial life and human-animal hybrids and our undefined moral responsibilities towards them. Of course much more can, and needs to, be said about these works of art, but these examples illustrate that works of techno-art provide rich material that requires hermeneutical, in-depth studies of artworks as well as normative-ethical reflection on the emotions and aesthetic and reflective experiences that works of techno-art give rise to, which can ultimately provide us with deeper insights into ethical aspects of new technologies.

Techno-Art and Public Deliberation
Works of techno-art can provide us with direct experiences with new technologies and probe emotional-moral reflection. They can hence play a powerful role in public deliberation. This is an idea that has not received much attention yet in the academic literature, and also not in practical methodologies for public deliberation on risky technologies.

There are numerous approaches to participatory technology assessment, aiming to provide for more democratic decision-making on risky technologies.[9] Works of techno-art could play a powerful role here. For example, works of art can play a crucial role in the debate on sustainable-energy technologies and on combatting climate change. They can help by making climate change more salient, inciting people to take actions,[10] and letting people critically reflect on the possible role of climate engineering. Artist Boo Chappel asked people to wear reflecting white hats, symbolising ideas from geo-engineering to protect the earth with a reflective layer to prevent global warming. Chappel then asked people to report and reflect about their very tangible, symbolic experiences and on the impact that such a technology would have. The works of Stelarc and Piccinini exemplify in a direct, experiential way the ambiguous feeling that many people have concerning developments in biotechnology. A common concern is that biotechnologists 'play with nature', but this is a complex notion that is difficult to conceptualise in a purely analytical way. Artists can play with the uncertainties and uneasy feelings by developing artworks via biotechnology and by examining the boundaries between life and technology. Such artworks may not necessarily provide for clear-cut answers, but they can provide for a common platform for reflection where different kinds of stakeholders can deliberate together on equal footing, as an artwork can make complex scientific developments concrete and tangible for a larger audience.

Another area where techno-art can make an important contribution is in critical societal reflection on robotics, AI and ICTs. Information technologies are deeply ingrained in our contemporary societies and many people due to the many conveniences that these technologies provide endorse specifically ICTs, such as computers and smartphones. However, ICTs can also lead to massive privacy intrusions. Increased automation may change our labour markets for good, by making large parts of society obsolete on the work floor. There are concerns about artificial intelligences getting out of control and eventually taking over from humans. That this is not only the material for vastly exaggerated science fiction novels and movies is exemplified by the fact that numerous leading scholars and technologists such as Stephen Hawking and Elon Musk have recently signed a letter warning against the dangers of AI and instead demand research on how to achieve 'beneficial artificial intelligence'.[11] Artists who work with artificial intelligence, robotics, and AI can play an important role in critical reflection on what it would mean for artificial intelligence to be beneficial, by exploring possibilities before they are introduced in

129

9 Cf. van Asselt & Rijkens-Klomp 2002.
10 Cf. Roeser 2012.
11 Future of Life Institute 2015.

society, but in more accessible, real-life settings than in the lab of scientists.

This way, techno-art can contribute to overcoming the so-called Collingridge dilemma.[12] According to this dilemma, the possibly detrimental effects of technologies can only be fully understood once these technologies are used by society, but then regulation might be too late and some may have already fallen victim to negative side effects. On the other hand, preventively restricting technology by regulation before it is part of society might lead to wrong estimates and forego society of possibly useful applications. Ibo van de Poel has proposed to see technology as a social experiment with explicit rules for aborting the process in case well-grounded ethical concerns arise.[13] However, even then undesirable, irreversible effects might already have taken place. Techno-art provides for an additional route before the introduction of technologies into society on a large scale. Techno-artists can go beyond the limitations of the labs of scientists and technology developers. They can take technologies to greater extremes and explore different scenarios in more tangible ways, and their works can be more accessible to society.

Challenges for Techno-Art
Techno-art can potentially make an important contribution to public debates, by highlighting societal implications, emotions and values related to new technologies. Artworks can inspire emotional-moral reflection on risky technologies. However, techno-art can also fail to do so. For example, artworks might be hard to grasp, appear manipulative or focus on unrealistic scenarios. Nevertheless, even in those cases artworks might function as a trigger for emotional reflection, deliberation and discussion.

Another concern is that while a work of techno-art might be helpful for emotional-ethical reflection, it might not be clear what its artistic and aesthetic merits are. For example, some people debate whether and in what respect Stelarc's project should be considered art.

Furthermore, there is a possible tension between artistic freedoms on the one hand and to explicitly invite artists to play a role in public deliberation on risky technologies on the other. It can be difficult to ensure that artist have the freedom to pursue their own ideas, while at the same time making a meaningful contribution to ethical reflection and doing justice to scientific and legal constraints.

Another question is whether artists should be bound by the same ethical and legal constraints as scientists and technology developers. As artists, they might arguably deserve more freedom, also because their works will probably not be produced on a large scale. On the other hand, while technologies might contribute directly to societal wellbeing, this might be less evident in the case of artworks, therefore making them less useful and, consequently, not worthy of the same room to introduce possible risks. Yet another position could be that artists and scientists deserve the same room for exploration without endangering the public, hence having to observe the same ethical and legal restrictions. There are as of yet no guidelines for these issues, but given the fact that there are more and more artists involved with risky technologies, these are all fascinating questions requiring further research.

The Promise of Techno-Art
To conclude, I think that techno-art can potentially make a constructive contribution to emotional-moral reflection on the risks of technological developments. Techno-art can help people to make abstract problems concrete, explore new scenarios, challenge their imaginations, and broaden their narrow personal perspectives through empathy, sympathy and compassion. Techno-art can contribute to an emotional-moral reflection about the kind of society we might want to live in. This means that techno-art can potentially contribute to public debate and overcoming stalemates. At the same time, there are various challenges for techno-art, such as how to preserve the non-instrumental nature of art while playing a role in public debates, and how to do so in a meaningful way. Techno-art is a rapidly expanding field of artistic development that could potentially make a major contribution to society, but which also requires thorough investigations by philosophers and other scholars.

12 Collingridge 1980.
13 Van der Poel 2013.

References

Adorno, T., Benjamin, W., Bloch, E., Brecht, B., Lukacs, G. 1980.
 Aesthetics and Politics. New York: Verso.

Asimov, Isaac. 1950. *I, Robot.* New York: Doubleday & Company.

Carroll, N. 2001. *Beyond Aesthetics: Philosophical Essays.* Cambridge:
 Cambridge University Press.

Collingridge, D. 1980. *The Social Control of Technology.* New York: St. Martin's Press;
 London: Pinter.

Gessert, G. 2003. Notes on the Art of Plant Breeding. *L'Art Biotech Catalogue,*
 exhibition catalog, Nantes: Le Lieu Unique, 47.

Goriss-Hunter, A. 2004. Slippery mutants perform and wink at maternal insurrections:
 Patricia Piccinini's monstrous cute. *Continuum: Journal of Media & Cultural Studies*
 18: 541-553.

Future of Life Institute. 2015. Accesssed at: *futureoflife.org/ai-open-letter/*

Gaut, B. 2007. *Art, Emotion and Ethics.* Oxford: Oxford University Press.

Harraway, D. 2007. Speculative Fabulations for Technoculture's Generations:
 Taking Care of Unexpected Country Originally published: (tender) creature exhibition
 catalogue. Artium. Accessed 5 August 2016 at: www.patriciapiccinini.net/writing/30/280/112.

Kompridis, N. (Ed.). 2014. *The Aesthetic Turn in Political Thought.*
 London: Bloomsbury Academic.

Nussbaum, M. 2001. *Upheavals of Thought.* Cambridge: Cambridge University Press.

Reichle, I. 2009. *Art in the Age of Technoscience: Genetic Engineering, Robotics,*
 and Artificial Life in Contemporary Art. Vienna; New York: Springer.

Roeser, S. 2012. Risk Communication, Public Engagement, and Climate Change: A Role for
 Emotions. *Risk Analysis* 32, 1033-1040.

Rorty, R. 1989. *Irony, Contingency, and Solidarity.* Cambridge: Cambridge University Press

Van Asselt, M. B., & Rijkens-Klomp, N. 2002. A look in the mirror: reflection on
 participation in Integrated Assessment from a methodological perspective.
 Global Environmental Change. 12:3, 167-184.

Van de Poel. 2013. Why New Technologies Should be Conceived as Social Experiments.
 Ethics, Policy & Environment. 16: 352-355.

Zwijnenberg, R. 2009. Preface, in Reichle, I. 2009. *Art in the Age of Technoscience: Genetic*
 Engineering, Robotics, and Artificial Life in Contemporary Art. Vienna; New York:
 Springer, xiii-xxix.

TECHNO-ANIMISM
WHEN TECHNOLOGY TALKS BACK

Jelte Timmer

You seem like a person, but you are just a voice in a computer.
- I can understand how the limited perspective of an unartificial mind
might perceive it that way. You'll get used to it.

Her[1]

In the 2013 movie *Her*, directed by Spike Jonze, a man falls in love with an intelligent operating system called Samantha. This operating system does not take on a physical form, like a robot, but exists only as a virtual presence, personified by a female voice. In the movie, we see how the protagonist develops a romantic relationship with the computer program Samantha. The themes addressed in the film are hardly novel: stories about humans who fall in love with artificial creatures go all the way back to Greek mythology.[2] Yet the movie also reflects a fundamental shift in how humans and computers interact with each other: the emergence of voice-guided interaction based on a largely invisible technology.

Since the dawn of computing, developers have been trying to come up with ways for us to interact with computers using our voice. It has been a process of trial and error, and only in recent years have conversational technologies started to take off. With applications such as Siri, Google Now, and Cortana readily available, most people have easy access to their own voice-enabled personal assistant. In 2015 Amazon launched Echo, a voice-enabled wireless speaker for home use, which you can ask questions and that improves as you get more acquainted. This is how a writer for technology monthly *Wired* describes his experiences with the personal assistant 'Alexa':

> Alexa quickly grew smarter and better. It got familiar with my voice, learned funnier jokes [..] In just the seven months between its initial beta launch and its public release in 2015, Alexa went from cute but infuriating to genuinely, consistently useful. I got to know it, and it got to know me. [..] This gets at a deeper truth about conversational tech: You only discover its capabilities in the course of a personal relationship with it.[3]

While Alexa does not have a fraction of Samantha's intelligence, the user is still able to develop a personal relationship with the device. The computer is given a name and a persona, and the style of communication is similar to how we interact with other humans. However, while the social aspects of the interaction are emphasised, the technology increasingly fades into the background. It is integrated with other devices we use or becomes part of our environment in the form of a sleekly designed object. How does this shift affect our relationship and interaction with technology? And, how are we to comprehend these forms of technology, which position themselves as social actors rather than as technological artefacts?

The End of the Computer and the Interface

A storied moment in the evolution of the personal computer is about the time when Steve Jobs visited the celebrated Xerox PARC R&D lab in Palo Alto, California in 1979 to view a demonstration of a new method of operating computers. Replacing the old, familiar model where users had to enter complex, text-based commands on a keyboard, researchers at PARC had created a computer with a device known as a 'mouse', which they used to click on icons, menus, and overlapping 'windows' on a screen. As legend has it, this was what inspired Jobs to design the

1 *Her*. Directed by Spike Jonze, 2013. Warner Bros. Pictures.
2 In the Greek myth of Pygmalion, a sculptor falls in love with one of his statues.

3 Pierce 2015.

Mac desktop computer, which popularised the graphical user interface we continue to use to this day.[4]

At the time of Jobs' visit, researchers in the same lab were busy developing another interaction model known as the 'conversational user interface'. Although the notion of people operating a computer by speaking to it has been around for a long time, we are currently seeing rapid advances in the development of conversational interfaces. This is both because computers have become faster and because software, aided by machine learning, can mine the massive amount of data related to speech and language that is generated on the Internet. Microsoft CEO Satya Nadella sees conversational interfaces as the next big thing in computing.[5] And recent research by Google revealed that the user of voice assistants such as Siri is popular among young people.[6]

The emergence of conversational technology coincides with a second trend, the trend that computers are receding into the background. In the 1990s, computer scientist Mark Weiser predicted that the computers of the future would no longer be the clunky grey boxes we had on our desks back then, but that they would decrease in size and be integrated into our personal environment.[7] Many companies were inspired in developing their vision of technology by what Weiser dubbed 'ubiquitous computing'. Philips launched the concept of 'ambient intelligence' in the late 1990s, which became a central concept in research funding from the European Commission. The idea behind ambient intelligence is to create smart technology that is integrated into the environment, adapts to the user, and anticipates the user's needs without requiring conscious mediation.[8]

Fast-forward roughly fifteen years; we have arrived at a point where these ideas have been developed into technologies that are starting to become a reality. As part of the 'Internet of Things', a growing number of products are being equipped with sensors, microprocessors, and Internet connections. Our everyday environments feature an increasing number of these types of smart devices, ranging from thermostats that automatically adjust temperatures to lighting and audio systems that adapt to our mood. The research and consultancy firm Gartner estimates that around 21 billion smart devices will be connected to the internet by 2020.[9]

133

The trend of computers being integrated into our environment signals the end of the computer as we know it: 'the death of general-purpose computing'. We are moving from working with one computer that – theoretically – can run any program, towards having different computers that are embedded into our environment and are each responsible for a specific type of program. The stationary computer that is controlled through a number of buttons will be replaced with smart systems in which the environment represents the interface and our behaviour serves as input.

Understanding 'Invisible' Computers

The move towards an Internet of Things, where computers are becoming less visible and integrated with our environments, makes it more difficult to understand exactly how computer systems in our environment work. Just like most of us do not know much about the intricacies of our nervous system, we also remain largely in the dark about the workings of a more unobtrusive computer system that automatically adapts to our behaviour. In this new situation, we also have to do without the trusty visual user interface that provides us with clues on how systems operate.

> It's a funny thing, trying to make sense of a technology that has no built-in visual interface. There's not much to look at, nothing to poke around inside of, nothing to scroll through, and no clear boundaries on what it can do.[10]

Various scientists have criticised the lack of transparency of so-called 'smart' environments.[11] In this new situation, technology is effectively a black box: it takes in input in the form of our behaviour and then makes specific decisions based on this input. For example, based on my

4 See note 1.
5 Weinberger 2016.
6 See note 1.
7 Weiser 1991.
8 Aarts and Marzano 2003.

9 Gartner 2015.
10 See note 1.
11 I put the word 'smart' in quotation marks here because the extent to which the technology can actually be described as 'smart' is debatable.

crammed schedule and the fact that I am speeding on the way home from the office, my 'smart' home environment concludes that I have had a stressful day and cancels all my appointments for the evening so I can kick back and relax. But then, the fast driving could also be because I am on a bit of a high after having had several successful meetings and am eager to share the news with others. Inaccurate analyses and false conclusions, or even outright manipulation will be hard to spot or correct for the average user. For incontrovertible proof of this, look no further than the recent Volkswagen emissions scandal, which involved the use of emissions test cheating software. Thus, how does our lack of understanding of the technologies we use impact the way we use them? As early as the 1970s, computer scientist Joseph Weizenbaum was puzzled by how people relate to technology they do not properly understand. Weizenbaum designed the renowned chatbot ELIZA, an automated chat program that, based on a number of pre-programmed scripts, made people feel they were actually interacting with a person. By processing users' responses to scripts and asking targeted follow-up questions, ELIZA was remarkably successful at enticing people into very detailed and personal conversations. Feedback from ELIZA users was extremely positive, with some psychologists even suggesting that the chatbot could be employed to perform some of the work of a psychotherapist. Weizenbaum was shocked at the very idea: 'ELIZA showed me more vividly than anything I had seen hitherto the enormous exaggerated attributions even a well-educated audience is capable of making, even strives to make, to a technology it does not understand.'[12]

When computers and visual interfaces disappear from sight, people will develop their own ideas of what their technological environment is made up of and how it functions. Weizenbaum posited that if people do not understand a certain technology, they will use the most obvious metaphor – their own intelligence – to create a concept of how it functions. As such, most people understand interaction with computer systems in terms of human behaviour and communication. This notion is supported by a series of experiments carried out in the 1990s by Stanford University researchers Clifford Nass and Byron Reeves.[13] They discovered that people showed the same behavioural patterns when interacting with media and computers as they did when interacting with other humans, particularly if the computer simulated forms of human interaction. For example, people had a more favourable opinion of computers if they apologised for making errors. Reeves and Nass formulated a number of statements about interaction between humans and various media, known as the 'Media Equation'. They argued, among other things, that people's perception of a device or application matters more than the actual technology behind it. The fact that a chatbot such as ELIZA comes across as an interested, inquisitive person is enough for users to engage with it as if it *were* an actual human being. This essentially eliminates the need for designers involved in artificial intelligence to create a system that possesses 'human-like' intelligence and understands human interaction, as long as this system creates an experience of intelligent communication and a sense of being understood.

Techno-Animism
Now, let us go back to Amazon's Echo and the new generation of 'smart' devices and 'smart' environments being developed worldwide. In an environment whose complexity is increasingly obscured, people are likely to come up with their own explanations of how the environment operates, and, in doing so, they will be guided by the types of interaction and thought processes familiar to them. The sociologist Nigel Trift has stated that we live in world where 'more and more things are able to become able'.[14] Washing machines decide themselves when to run a load and if it is time to re-order detergent. Or, when you arrive home after a rough day at the office, the smart home environment will play you a funny YouTube video while the barrage of notifications flooding your phone is temporarily muted to give you time to decompress.

Design theorist Betti Marenko has argued that the oblique nature of technology environments and their 'human' capacity to act cause users to understand devices in animistic terms. This means that objects are interpreted as if they were 'living' beings with their own personalities. Animism is an ancient spiritual concept that holds that objects such as stones, plants, rivers, or

12 Weizenbaum 1976: 4.
13 Reeves and Nass 1996.
14 Thrift 2011.

thunderstorms have a soul or spirit. It is found in many different cultures and may be one of the oldest ways of making sense of the world. The Swiss psychologist Jean Piaget discovered that many young children also use animistic interpretations when discussing the world around them, for example when they say the sun is hot because it likes to keep people warm.

In a complex environment, animistic interpretations help to create order in a world that is otherwise abstruse. Marenko writes: 'As this landscape of uber-connectivity becomes the invisible, intangible, and 'switched on' backdrop to our daily lives, we deploy an animist outlook to give meaning to an otherwise fairly incomprehensible world of objects.'[15] A complex technological environment, then, leads to the emergence of techno-animism as a mental model for discussing and relating to actions performed by the smart devices that surround us. What matters is not so much if we really do have animistic beliefs – i.e. if we actually believe a thermostat possesses intelligence and is alive – but how animism as a mental concept for describing the world is changing the way we interact with and relate to that world.

Marenko demonstrates that techno-animism is enhanced by the way modern technologies are designed. Since much of the functionality of smart devices originates in the digital world, the design of the object in the physical world becomes increasingly neutral and more standardised. The white rectangular box is a commonly used design format for smart devices, including smoke detectors, smart TV modules, and thermostats. The simple and often closed design belies the complexity behind these devices.

Interaction with these devices is also geared to human forms of social interaction, which obfuscates the complex technology underlying the device. A good example is the smoke detectors manufactured by NEST (see Figure 2) – the same company that created the popular 'smart' thermostat. Rather than simply sounding an alarm at full blast the moment a whiff of smoke is detected, these devices 'think before they talk'. A human voice tells the resident about the problem and where it is located, a script that might sound something like this: 'Watch out, carbon levels in the bedroom are very high. Get some fresh air now!' The tagline for the 'smart' home console Twine – which tells users when their laundry is done, how much time the kids spent watching TV, and whether anyone tried to break into the house while they were out – is: 'Listen to your home, wherever you are.' In this case, the home is presented as a smart, animated object that does your thinking for you, alerts you when something is wrong, and with which you can even engage in 'conversation' if the situation calls for it.

Voice-activated assistants on smartphones, such as Siri, have normalised the practice of talking *to* objects rather than *about* them. The role of technology is shifting from objects that mediate conversation to objects that are actually part of the conversation themselves. This form of interaction is forcing us to rethink the way we relate to objects – after all, we are essentially entering into a social relationship with them by talking to them. This makes Spike Jonze's movie *Her* more than a prescient vision of the future – it is also very much about how our relationship with technological objects is changing in the here and now.

A New Relationship between Humans and Technology

With a shift occurring from the command line to a conversational interface (see the figure below), we have come to interact with an *image* of technology rather than with the technology itself. The lines from the film *Her* quoted at the beginning of this text expose this conflict: 'You seem like a person but you are just a voice in a computer.' This represents the conflict between the image that is present and what the technology is in reality. We respond to the presence in the sense that we feel the person really exists, even though in reality we know we are deluding ourselves.

As we have seen, our relationship with technology is changing. You could say it is a shift from *using* technology to *interacting with* technology. As part of this interaction, we are using conversational interfaces to enter into a relationship with technology in a human, social way. This kind of 'human' interaction with technology has certain advantages. The use of ingrained human forms of interaction is an effective approach to design, which allows people to interact with their

15 Marenko 2014: 231.

environment in a way that is natural and intuitive. Regarding technology as a humanlike other may even help us to take a more respectful attitude towards our environment rather than viewing it as merely an object available for our convenience.

At the same time, an animistic perspective can also have an obfuscating effect by hiding the fact that technology is just that – technology: an artefact that was designed by a manufacturer based on a specific rationale. The fact that technology 'reads' and analyses us while we lack even basic understanding of that technology creates a power imbalance. The technology gains detailed knowledge of the user, while the user has no clue as to how this information is used and analysed. Designer and cognitive scientist Don Norman explains how this creates a situation of distrust between humans and their technological environment: 'The system's methods remain hidden so that even if [we] were tempted to trust it, the silence and secrecy promotes distrust.'[16] According to Xerox PARC researcher Mike Kuniavsky, the lack of understanding can lead people to resort to technological superstition when dealing with complex technological environments. He states that 'technology-based rituals and superstitions may occur as intelligent objects appear in ever-increasingly intimate situations'.[17] Examples of this would be someone waving their mobile phone in the air in an attempt to 'catch' signal reception or people going through the same ritual every time they switch on their computer – first the computer, then the monitor, then the printer – because they believe it is not going to work otherwise.

The hidden complexity of technology diminishes the user's understanding and autonomy if it results in the user bending to the 'whims' of the technology in question. An equitable relationship with technology would require the user to develop an understanding of the technology – just as the technology has come to understand them. If we are to adopt a sensible approach to a future generation of technology, we need to start thinking in terms of augmentation rather than automation. To create a future where human beings and technology work together instead of one imposing its scripts or 'will' on the other.

136

16 Norman 2007: 7.
17 Kuniavsky 2004: 2.

References

Aarts, E. and Marzano, S. 2003. *The New Everyday: Visions of Ambient Intelligence.*
    Rotterdam: 010 Publishers.

Gartner, P. 2015. 'Gartner Says 6.4 Billion Connected "Things" Will Be in Use in 2016,
    Up 30 percent From 2015'. Press Release. Accessed at:
    www.gartner.com/newsroom/id/3165317.

Her. Directed by Spike Jonze, 2013. Warner Bros. Pictures.

Kuniavsky, M. 2004. Animist User Expectations in a Ubicomp World: a position paper for
    Lost in Ambient Intelligence. Proceedings of CHI. ACM Press.

Marenko, B. 2014. Neo-Animism and Design. A New Paradigm. *Object Theory.*
    Design and Culture. 6:2.

Norman, D. 2007. *The Design of Future Things.* New York: Basic Books.

Pierce, D. 2015. We're on the Brink of a Revolution in Crazy-Smart Digital Assistants.
    WIRED, 16 September 2015. Accessed at: *www.wired.com/2015/09/voice-interface-ios/#slide-2.*

Reeves, B., Nass, C. 1996. The Media Equation: How People Treat Computers, Television,
    and New Media like Real People and Places. Cambridge, MA:
    Cambridge University Press.

Thrift, N. 2011. Lifeworld Inc. – and What to do About it. *Environment and Planning D:
    Society and Space,* 29:1, 5–26.

Weizenbaum, J. 1976. *Computer Power and Human Reason: From Judgment to Calculation.*
    San Francisco: W.H. Freeman and company.

Weiser, M. 1991. The Computer for the 21st Century. *Scientific American,* 265:3, 66-75.

Weinberger, M. 2016. Satya Nadella says Microsoft's next big thing will have 'as profound an
    impact' as touchscreens and the web. *Business Insider,* 30 March 2016. Available at:
    *uk.businessinsider.com/microsoft-ceo-satya-nadella-on-conversations-as-a-platform-and-chatbots-
    2016-3?r=US&IR=T.*

DESIGN OF WISDOM COACHES FOR END-OF-LIFE DISCUSSIONS
MIXED REALITY, COMPLEXITY, MORALITY, AND NORMATIVITY

Maarten J. Verkerk[1]

Under influence of modern technology, the face of Western healthcare changes radically. The application of new medical technologies in combination with genetic information and Big Data disruptively transform the whole healthcare chain. A promising area of new healthcare technologies is the field of augmented reality, virtual reality, and gamification. In this field some aspects of reality are augmented, other aspects are virtualised, and in some cases gaming concepts are applied. In this article the fields of augmented reality and virtual reality will be called 'mixed realities'.

The phenomenon of mixed realities is not a futuristic invention of science fiction, but mixed reality is already used in today's healthcare; for example, nurses can find veins more easily with a mixed reality approach, by using a handheld scanner they can determine the position of the veins in the arm or hand of the patient. This information is converted to an image of the veins and projected on the skin of the patient. This application makes it easier to succeed piercing the vein on a first try. Another application is the use of Google Glass to support young mothers in breastfeeding. Mothers receive step-by-step instructions as they begin learning to breastfeed. Additionally, they can video call a nurse or counsellor who can view the process through the Glass's camera and give the mother immediate feedback and suggestions. Another important application of mixed realities is the field of surgery. Without any doubt, we stand on the eve of a revolutionary development that will influence every aspect of human life.

The development of mixed realities in the field of healthcare raises many moral and philosophical questions. Such as, what is the status of mixed realities? How do humans relate to mixed realities? What moral questions and objections arise? How to relate the aesthetic aspect to the healthcare aspect? How are mixed realities embedded in social relationships? And so on.

In this chapter we will investigate these types of questions in view of the development of a wisdom coach that supports frail elderly and/or their beloved to address emotions, moral problems, and existential questions at the end of life.

What is a Mixed Reality?

The idea of a mixed reality is visualised in many science fiction movies. For example, in *The Matrix* both the real and virtual world are present and in *Minority Report* an augmented reality supports the officials to prevent crimes. Paul Milgram et al. define a mixed reality as an intermediate state on the continuum of the real world and the virtual world.[2] Human beings experience the real world directly or indirectly (for example, through monitors and phones). One pole of this continuum is the real world that exists solely of real objects, such as humans, animals, trees, cars, houses, tools, and so on. The other pole of this continuum is the virtual world that consists solely of virtual objects that can be experienced by monitors, phones, or immersives. An immersive is a system that generates a three-dimensional image that appears to surround the user. An augmented reality is always a mediated reality and therefore a 'reduced' reality. We experience the real reality with all our senses. However, mixed realities are mediated by technological artefacts and they appeal to a limited number of senses, mostly the eye and the ear. Finally, mixed realities are always embodied in technological artefacts whereas the senses are embodied in a biological body.

Within this framework it is straightforward to define a mixed reality environment as one in which real world and virtual world objects are presented together within a single display, that is, anywhere between the extrema of the real-virtual continuum. We speak about an augmented

138

1 I would like to thank Teun Aalbers, Karin Alfenaar, Rozemarijn van Bruchem, Jan Jonk, Willem-Jan Renger, Lavender She, Jan Peter de Vries, Michiel van Well, and Esmé Wiegman for the stimulating discussions.
2 Paul Milgram et al. 1994.

reality when this mixture is more to the real pole and about an augmented virtual world when it is more to the virtual pole, see Figure 1 From these definitions augmented reality can be defined as a live or direct view on the real world, which is augmented by computer-generated input such as video, graphics, verbal instructions, sound, or GPS-data. Presently, augmented reality is embedded in monitors, phones, and so on. In the future it could be integrated in glasses, headsets, or digital contact lenses. A similar definition can be formulated for augmented virtual worlds.

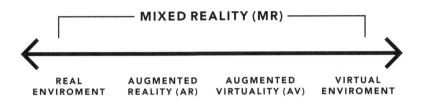

139 · Figure 1: Real – Virtual continuum in according to Milgram et al. (1994).

The Meaning of Dying

A frequently asked question is 'what is the meaning of life?' Yet hardly anyone in Western society asks the question what is, or can be, the meaning of *dying*. More often than not, dying is seen an unpleasant consequence of living, it makes the living uncomfortable and at a loss for words, thus discussions about 'the end' are preferably avoided. Moreover, we do not believe that dying has any meaning so the expressions 'meaning of dying' is experienced as a contradiction in terminus. In practice, we hide death behind white doors and patients die in loneliness.[3] The question of the meaning of dying is an existential question. It touches our deepest beliefs about what it means to be human, a mortal human. While we try to add meaning to our lives we rarely think about how we would like to die, what we need to say, achieve, or complete before we feel at peace with our approaching departure.

The meaning of dying concerns taking leave of beloved, reflecting on our lives, but also finding comfort in each other, both the deceasing and the living. The question regarding the meaning of dying also can be called a 'slow question'. That means that it takes time to address and to readdress this question. Every person involved in this process will have a different timeline.

These matters and questions should be taken into consideration by the wisdom coach who guides the end-of-life discussion.

End-Of-Life Discussions

Despite, or because of, the feeling of discomfort, end-of-life discussions that revolve around the meaning of dying have become increasingly important in healthcare. One of the reasons is that the possibilities of modern medical technologies exceed the bodily capacities of frail patients. In present-day healthcare, often (life-prolonging) treatments are given of which the negative side effects outweigh the health benefits.[4] These treatments – often indicated as 'overtreatments' – lead to an increase or lengthening of the patient's suffering.

Boer et al. have conducted an exploratory investigation on the phenomenon of over-treatment.[5] They propose two approaches to deliver appropriate care and to prevent overtreatment. First, at the core of the end-of-life discussions should be the meaning of dying and *not* medical technology. Second, three supporting values become apparent after we place the meaning of dying at the centre of the discussions:

1. Respect for autonomy: optimal consideration for the patient's wishes;
2. Non-maleficence: refrain from doing harm;
3. Beneficence: make every effort to promote people's wellbeing, including protecting their lives.

3 Elias 1985.
4 Kaufman 2005; Boer et al. 2013.
5 Boer et al. 2013.

A challenging question is: how to implement this approach – from now on indicated as the meaning approach – in healthcare? This question can be divided in two subquestions. First, what are the areas where healthcare and frail elderly (and their beloved) 'touch' each other? This has to be established to foster the assumption of the meaning-approach. Second, how can we embody this approach? In this article I address only the idea of a mixed reality approach complemented with gaming principles.

Explorations in Designing a Wisdom Coach
A wisdom coach who guides the end-of-life discussions has to cope with plurality with respect to a variety of worldviews and religions. Can a wisdom coach address the whole field? Or should the coach limit himself to a first exploration and then hand the care to pastors, priests, imams, or other coaches for specific support? The aesthetic design (or, the presentation and visibility) of the wisdom coach appears to be very critical. On the one hand, the design has to be nearly absent from the scene in order to prevent people from being too aware and, possibly, scared off by the application. On the other hand the visibility and presentation of the coach has a large invitational effect and holds the key to a successful trajectory. Furthermore, dying is a social process that involves family and friends. The social aspect raises many questions. How do the different people involved interact? Can the coach repair broken dialogues? Are relatives and caretakers allowed to continue to ask the patient questions (via the coach) even when he has indicated he does not wish to answer them?

How should the wisdom coach address the matters of autonomy, non-maleficence, and beneficence? Is the coach allowed to raise questions? For example, when all parties involved focus on medical questions, is the system allowed to raise existential questions?

All these considerations show that the design of a wisdom coach is a big challenge. The coach needs to be very 'empathic' with respect to the mood of every participant and has to address questions that are often not asked. For that reason, it is expected that principles of mixed reality and gamification could be suitable approaches. Figure 2 gives an example of a design: a life tree whose size and shape changes under influence of all dialogues and a background that changes depending on the mood of the user. The users can ask each other questions by writing a letter and pinning it to a virtual tree, which can then be read by the other(s).

Philosophical Analysis
The first exploration of the role of the wisdom coach clearly shows that it is a complex matter. At the very least, religious, aesthetic, social, moral, and medical aspects play a role in troubling the waters. This observation raises the question (1) whether this list of aspects is complete, (2) how these aspects relate to each other, and (3) how to prevent one aspect dominating the others. To address these questions I would like to make use of the work of the Dutch philosopher Herman Dooyeweerd (1894-1977).[6]

Dooyeweerd makes a distinction between 'wholes', 'aspects' and 'dimensions'.[7] A 'whole' is a complete 'unity', 'system', or 'entity' with an own character or identity. Examples of wholes are human beings and animals, trees and bushes, stones and grains of sand. All these wholes are present in the natural environment and have an own identity. Technological artefacts are also wholes; examples are monitors, phones, hospitals, and churches. These artefacts also have different functions and have an own identity. So, our first conclusion is that there are different kinds of wholes and every whole has its own identity.[8]

Dooyeweerd showed in his theory of modal aspects that 'wholes' function in a number of different aspects or dimensions. For example, a human being needs food (biological dimension), perceives the environment (sensitive dimension), interacts with other people (social dimension), buy goods in a shop (economical dimension), enjoys art (aesthetical dimension), shows ethical behaviour (moral dimension), and does or does not believe in a transcendent God (spiritual or religious dimension). Dooyeweerd distinguishes fifteen different modes, see Figure 3. By means of an in-depth philosophical analysis he argues that all these dimensions have their own character that

6 His concepts regarding technological artefacts have been used by
Verkerk et al. 2016.
7 Dooyeweerd 1969.
8 Verkerk et al. 2016: 89-105.

expresses itself in its own dynamics, mechanisms, and laws or norms. For example, the nature of the economic aspect is quite different from the nature of the spiritual aspect. The dynamics of the social interaction are quite different from the subtleties of enjoying art. Finally, the biological laws that determine the digestion of food are quite different from the norms for moral behaviour. In other words, every dimension has its own nature and cannot be reduced to other dimensions.[9]

Technological artefacts are also wholes that function in different aspects. For example, a wisdom coach has a physical dimension (the mediating technology consists of materials with specific properties) and a social dimension (facilitates social interactions between users). It also functions in the economic dimension (price) and in the aesthetic dimension (beauty of the design of mixed reality). It also functions in the juridical dimension (intellectual property) and in the spiritual dimension (providing hope and trust).[10]

Spiritual: devotion, trust, transcendence
Moral: love, care
Juridical: justice, guilt
Esthetical: beauty, harmony
Economic: scarcity, exchange
Social: interaction, intercourse
Linguistic: symbolic meaning
Power: shaping, influence
Logical: analytical, rational
Psychological: perception, emotions
Biotic: growth, decay, reproduction
Physical: energy, interaction, process
Kinematical: movement
Spatial: space
Numerical: quantity

141 · Figure 3: The fifteen modal aspects as proposed by the philosopher Dooyeweerd (1969).

The theory of modal aspects can be used to refine our considerations on the identity of technological artefacts. Let us compare two different electronic coaches: a coach to choose the cheapest airplane tickets and a wisdom coach. The design of a coach for airplane tickets can only be understood from its economic function. The whole website is designed in such a way that users are informed about prices and conditions and are seduced to buy a ticket via this website. Here, the economical aspect determines the identity of a coach for airplane tickets. The design of a wisdom coach only can be understood from its spiritual or religious function: to address existential questions, to contemplate the meaning of dying, and to comfort each other. Here, the spiritual or religious dimension determines the identity of a wisdom coach.[11] The aspect or dimension that characterises the identity of an artefact is called the 'qualifying aspect'.

This qualifying function plays a leading role in the design of technological artefacts; or rather, the spiritual or religious function of the wisdom coach guides, or discloses, all other aspects. For example, the designer is not at full liberty to design his aesthetic position but this aspect has to be carefully designed in such a way that it serves its existential function. The same thought applies to the social aspect; the interaction of all users has to be designed in such a way that end-of-life discussions will happen.

To summarise, we started by establishing that a wisdom coach is a complex system. Our philosophical analysis shows that a wisdom coach is characterised by a total of fifteen aspects. All these aspects, which have their own nature and character, have to be taken into account in the design process. Additionally, this analysis revealed the importance of the so-called qualifying function that leads or directs the design of the artefact. In the case of wisdom coaches, the spiritual or religious aspect has to guide or disclose all other aspects.

Mixed Conclusions
Mixed realities combined with genetic information and big data will induce radical changes in healthcare. This development raises many moral and philosophical questions. These questions are

9 Verkerk et al. 2016: 62-85.
10 Verkerk et al. 2016: 89-105.
11 Verkerk et al. 2016: 89-105.

investigated in view of the development of a wisdom coach for end-of-life dialogues. The follow-ing conclusions are drawn:

1. Mixed realities are always mediated realities.
2. Mixed realities are complex. Its complexity can be unravelled by the philosophy developed by Herman Dooyeweerd. It is shown that a complete description has to address the fifteen different aspects and that each aspect has its own nature and normativity.
3. The design of mixed realities in end-of-life discussions has to be guided by the spiritual or religious aspect.

References

Beauchamp, T.L. and Childress, J.F. 2009. *Principles of biomedical ethics.* 6. Oxford: Oxford University Press.

Boer, T., Verkerk, M.J., and Bakker, D.J. 2013. *Over(-)behandelen. Ethiek van de zorg voor kwetsbare ouderen.* (Over(-)treating. Ethics of care for frail elderly – in Dutch). Amsterdam: Reed Business.

Dooyeweerd, H. 1969. A new critique of theoretical thought. 4. Philadelphia PA: The Presbyterian and Reformed Publishing Company.

Elias, N. 1985. *Loneliness of the Dying.* New York: Blackwell.

Kaufman, S.R. 2005. *And a time to die. How American hospitals shape the end of life.* New York: Lisa Drew.

Milgram, P., Takemura, H., Utsumi, A., and Kishino, F. 1994. Augmented Reality: A class of displays on the reality-virtuality continuum. *Proceedings of Telemanipulator and Telepresence Technologies SPIE.* 2351, 282-292.

Verkerk, M.J., Hoogland, J., van der Stoep, J., and de Vries, M.J. 2016. *Philosophy of Technology: An Introduction for Technology and Business Students.* London: Routledge.